728
Sh 44k

From *Trattoro di Architettura*, 1464, Antonio Averlino Filarete.

1

Contents

Shelter II

Distributed in the United States by Random House and in Canada by Random House of Canada, Ltd.

Individual copies of this book available by mail; see p. 222 for details.

Shelter Publications
P.O. Box 279
Bolinas, Calif. 94924 USA

Shelter Publications is a non-profit educational corporation formed for the purposes of providing research, design and education in the fields of housing and the building crafts; cataloguing and preserving traditional as well as innovative construction techniques; maintaining a network of contributors; and disseminating information to the public by publication of directly related literature.

Library of Congress Cataloguing In Publication Data
Main entry under title:

Shelter II

    1. Dwellings 2. House Construction. I. Kahn, Lloyd.
TH4812.S53   728   78-57133
ISBN 0-394-50219-1
ISBN 0-394-73611-7 pbk.

We are grateful to authors and publishers of the following books for permission to reprint copyright material:
CALIFORNIA LIVING MAGAZINE of the San Francisco Sunday Examiner and Chronicle, September 19, 1976; copyright (c) 1976, San Francisco Examiner. Reprinted with permission.
FUNDAMENTALS OF CARPENTRY, VOLUME 2, PRACTICAL CONSTRUCTION (Third Edition) by Walter E. Durbahn and Elmer W. Sundberg; copyright (c) 1964 by American Technical Society. Reprinted with permission.
PIONEER TEXAS BUILDINGS, Clovis Heimsath. University of Texas Press, 1968; copyright (c) 1968 by Clovis Heimsath.
PRIMITIVE ARCHITECTURE AND CLIMATE by James Marston Fitch and Daniel P. Branch; copyright (c) December 1960 by Scientific American. Reprinted with permission.
SCIENCE YEAR – THE WORLD BOOK SCIENCE ANNUAL; copyright (c) 1975 Field Enterprises Educational Corporation. Reprinted with permission.
A SMALL HOUSE IN THE SUN; copyright (c) 1971 by Samuel Chamberlain, permission by Hastings House, Publishers.
WORKING: PEOPLE TALK ABOUT WHAT THEY DO ALL DAY AND HOW THEY FEEL ABOUT WHAT THEY DO, by Studs Terkel; copyright (c) 1972, 1974 Studs Terkel. Reprinted by permission of Pantheon Books, a division of Random House, Inc.

A shepherd's refuge from the wind, Aksehir, Turkey, 1978.

Introduction

Shelter II is the second in a series of books about people building their own homes in different parts of the world. *Shelter*, a scrapbook of building ideas, was published five years ago and since then, housing costs — land, building materials, real estate, rents — have increased dramatically. The principles outlined in *Shelter* seem even more important today: re-learning the still-usable skills of the past, finding a balance between what we can produce for ourselves and what we must buy, and doing more hand work in providing life's necessities. *Shelter II* goes on with a review of world-wide housing techniques, provides a basic manual of design and construction for the first-time house-builder, and covers self-help housing projects now underway in large cities.

The book begins with simple shelters still being built and lived in by people with minimal resources. They can be viewed either for historical or anthropological interest, or as sensible — even instructive — examples of efficient construction by those who lack the choices available in industrialized societies. We can also learn from the farm and country buildings of North America — still-standing reminders of an era of practical design and straight-forward construction practices: siting to minimize wind exposure, roofs shaped to shed rain or snow, shady porches for summer coolness.

Stud framing has been the most common housebuilding technique in this country since sawmills began turning out 2 x 4's and 2 x 6's in the mid-nineteenth century and is shown next as the most practical form of house construction in most situations today. There is an introduction to the principles of design, framing drawings of seven roof shapes, and a 24-page abbreviated construction manual for building a small home.

In some cities, abandoned buildings are being cleaned out and rehabilitated, older houses repaired and maintained. People are working to create their own living space and learning new skills in the process; derelict neighborhoods are revitalized, and housing is provided where it is needed most. Some more recent developments are also examined: dome housing, and America's current program to establish colonies in space are reviewed and commented upon.

There are also personal accounts and seasoned advice from builders in different climates, with a variety of design approaches, construction techniques, and building materials: adobe in New Mexico, log cabins in Washington and Idaho, a family-built stone house in California, homesteading on a Scottish island, floating summer tents in Alaska, and houseboats in Amsterdam.

Throughout, there are consistent elements. Practical builders, wherever they live, work with simple techniques and what is most readily at hand: earth, thatch, stone, milled lumber or abandoned city buildings. Weather, purpose, materials govern design. Tradition, experience, practice determine building technique. Individual initiative and hand labor by owners can decrease spiraling costs and reduce or eliminate life-time mortgage obligations.

In the past century, industrial-technological progress has been rapid. Yet basic human needs are still much the same. Shelter has always meant a roof overhead, protection from the elements, a refuge. A home is still a place for working, resting, sharing, healing, dreaming . . . some things haven't changed that much.

3

At left: *cliffside residence, Cuenca, Spain.* Below: *fish nets drying on the island of Janitzio, Lake Patzacuaro, Mexico.* Bottom: *stone house under construction in Nepal.*

Indigenous Builders

In much of the world, homes are still built with a minimum of machinery, electrical energy, and processed materials. Mud, woven grass and straw in Africa; reeds, ivory nut leaf, bamboo in the South Pacific; stone, mortar and whitewash in Greece. No plywood, sliding glass doors, butyl caulk or skilsaws.

People build this way when there is no other choice: the economics of necessity. These buildings blend with their surroundings because they are products of those same surroundings. Design is based upon close observation of weather patterns, changing seasons, and available materials. Construction is often a community affair, with tools and techniques worked out over generations.

There is much to learn from indigenous builders, past and present. Not that we will suddenly return to the bamboo grove, or begin building huts of mud and thatch. But in an era of diminishing resources and rising prices the "primitive" solutions of indigenous builders can be instructive in planning to provide our own shelter needs — having less resources requires greater resourcefulness: breezes, rather than air conditioners for summer cooling; sensible siting to minimize wind exposure; rock that is cleared from the fields for farming used to build the walls of the farmhouse; design based upon available materials and minimal consumption rather than abstract concepts or architectural frivolity.

On the following fifty pages are photos, drawings, and descriptions of low-cost, energy-efficient structures from around the world, from the Nabdam family compounds of Ghana to the tule mat lodges of the Mexican Kickapoo. Homes built with few of the benefits of mechanical industry, yet with great economy, skill and beauty.

At left: *farmer's house, Lambwe Valley, Kenya, Africa.* Below: *thatched roof interior of hotel in Twiga, Kenya, Africa; approximately 90 foot diameter.*

Ghana

by Paul Marchant

David is a Nabdam tribesman, a native of Ghana. We met in the silent white-hot glare of late November's noonday sun. The heat fanned by the dust-laden Harmattan wind prematurely withered all life before it. David stood in a group of boys and old men who waited with suspicious eyes while I explained my interest in their picture decorated houses. I presented a customary gift of cola nuts wrapped in a leaf, asking permission to visit their compound and photograph the beautifully drawn crocodiles and birds. The expressions of doubt which had contorted these aged yet youthful faces of the earth began to ease. As we started to move around the compound David became the vehicle of an oral tradition unbroken for at least 500 years, and began to describe the remote origins of his people.

In olden times there was once a fairy who lived in a thick grove of trees. This gave him shelter and protection as he didn't know how to build a house. The fairy was a farmer with great knowledge of the plant and animal kingdom, and had supreme contact with the earth spirits.

One day a hunter emerged from the bush, exhausted after the long pursuit of game.

He 'eyed' the fairy's activities with great surprise, unable to believe that anyone could survive by eating grass-seed and plants.

The fairy explained his craft and the hunter was so pleased that he implored the fairy to teach his villagers.

The fairy went with him and spent a year in the village so that the people would know the complete farming cycle.

The hunter was delighted and gave the fairy his youngest daughter for a wife.

The fairy returned to the grove and his new wife showed him how to build a house and live in it with a wife.

Their sons and daughters became the Tendaanas (High Priests) of the Land....

David asked me if I wanted to watch his uncle — a soothsayer — beginning work on a new house, across the fields. We would visit his father, Kapeon's house on the way .... We picked our way along a narrow time-trodden path. On either side, spiky stubble protruded from the rippling heaps of soil moving away from a compound at their centre. A few weeks before, guinea-corn and millet had been a vivid green barrier ten feet high, obscuring the houses and small trees. Now the strewn remains of the crops burned yellow and brown in the strong heat, in the mid-nineties during the dry season.

Kapeon's house was typical of Nabdam family compounds, consisting of a circular chain of cylindrical cells measuring eight or nine feet in diameter and connected by screen walls. A single entrance opened into the internal livestock yard, separated from the living courtyards and sleeping rooms at the far end by a major dividing wall containing the granary. Indigenous materials make optimum use of environmental potential: all walls were laterite

Wall at which dangerous food is cooked.

Lamsin   **David with friends.**

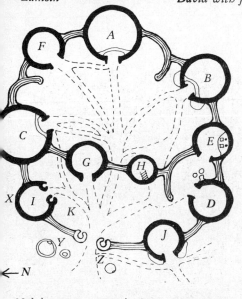

Nabdam compound near Zuaraungu. About 50 feet in diameter. A. compound leader; B. wife; C. children; D. kitchen; E. grinding room; F. store; G. fowls; H. granary; I. goats. J. guard room; K. hen coops; X. ancestors' door; Y. & Z. ancestors' shrines.

(red earth) rendered with cow dung/clay; roofs were conical thatches or flat; even household implements (including calabash dinner plates and leaf food wrappings) were biodegradable, and wastage is comparatively unknown.

David's room was the only rectilinear cell in the compound; he told me the old men build these rooms for their senior sons. Kapeon and his children, who

live with him in the house, form the basic social unit of the Nabdam, the minimal lineage. The maximum lineage, which is the basis of clan structure, consists of men and women descended in an unbroken line from a single male ancestor. The clan unites one or more families in the worship of common ancestors, symbolised as an animal totem in their myths of origin.

At the end of a five-minute walk from Kapeon's house, we entered a clearing and greeted David's uncle Lamsin, the soothsayer, at the new house sitc. He was a tall lean figure with well-formed features framing wise, bright eyes. He and his helpers were completing a course of the walls. Two courses were laid, one at the beginning and one at the end of the day. It was now approaching siesta-time and the builders would soon retreat to the tree shade of a local beer bar till the evening cool . . . .

I asked David, "What does a man have to do in order to acquire land and start a house?" With the help of an Elder, David related that a man . . . is eligible to build a house if he is the first born of a householder and has a wife with children. The implementation of this right constitutes initiation into an advanced state of manhood.

Lamsin first acquired land from the Tendaana, which was confirmed by the chief. He then began the preliminary rites by going to visit the soothsayer with an elder (even though he was a soothsayer himself). The soothsayer divined whether it was propitious for him to erect the dwelling, and the ancestors

answered to grant permission.

The second step was to clear the site. It is customary for friends and relatives to help the new house-owner, their numbers depending on his good reputation. Lamsin led a large work force of 20-30 men. Certain relatives must help on the first day of building, after which all helpers are rewarded with food and Pito (local beer fermented from guinea-corn).

Elders were informed about the day of gathering at the new house site; once they were together in their circle, the third step was to "try out" a fowl. A pot contained the roots of certain auspicious trees and water. A shallow dome lid was fitted creating a rough sphere. The bird was then sacrificed over this round. Its blood and small breast feathers were smeared on the lid and the body cast groundwards for its death struggle. The encircling audience waited in anticipation for its final resting position. If the fowl came to rest facing the sky then God and the ancestors were pleased and it was the correct site. If the fowl lay face downwards the site was wrong. The sky is synonymous with the patrilineal clan and the earth or "blood" with the female clan.

Lamsin got a negative result from the first trial and hurriedly consulted another soothsayer. He was told that his father, who had been a powerful soothsayer before him, still had a living influence over the village although he died in 1968 . . . . The soothsayer concluded that the JuJu must be exhibited at the new house site so that the people would see who was responsible for Lamsin's

continued

move to the new building. This was carried out and another fowl was sacrificed, producing the desired result.

A day was then selected for the women to fetch water for making mud bricks used in wall construction. The women started carrying water on the fourth day after the gathering of the Elders. The young men assisted in mixing the soil, which was left in mounds for the following day. On the day when house building commenced, the young men came out early in the morning to start turning the soil in preparation for the arrival of the Elders. The Tendaana accompanied them and laid the foundation by making his mark on the ground. He called on God and the ancestors, asking for a good beginning and a good end to the building and long and healthy lives for its occupants.

The mason (in this case Lamsin, or another mason if he had been unskilled) marked the plan of the first circular cell: without a compass he 'dances' in a circle furrowing the ground with his toes. He made apparently perfect circular plans: when measured the plans were highly accurate. The Elders debated the size and position of the room and the plan was erased and danced again until agreement was reached. Lamsin's room was first to be built, the next was for his wife, followed by those for his children and animals.

The granary, the most vital cell (its grain contents fluctuate from sufficiency to starvation level) symbolised the unity of the family. It was also the most beautifully formed cell, utilising material of superior elasticity — special clay mixed with cow dung, straw and okra sap.

Before the "wet" bricks were laid for the first room, Lamsin sprinkled protective water from the sacrificial pot into the furrowed circle. He and his helpers formed round lumps from the mixed laterite mounds positioning them round the circle. They began at about a foot in diameter and decreased with each subsequent course. The wall tapered from a foot at the base to between four and six inches at the top. The courses measured 12"-15". Lamsin showed that a pitched roof required six courses and a flat roof seven, producing walls between seven and eight feet high, kept vertical by hand and eye alone. When struck, after drying, they resonated with the pure timbre of a well made pot. The walls were plastered with a mixture of cow dung, clay and vegetable juices, worked with wooden trowels, then sized with an extract boiled from Dawa Dawa pods. The finish was a reddish-brown colour, providing the ground for wall decorations of animal proverbs. These paintings, executed in the local earth paints of red, white and black, are usually made by a man's wives in competition for his valued praises. The finished wall lasts between two and three years.

A pitched roof (average gradient 1:1½) was erected on the mud cylinder, first using a half octahedron constructed with four main rafters. Their forked ends were thrust into the drying mud at the eaves and bound at the crown with rope. Lamsin judged their position and filled in the conical frame work between the rafters of the original pyramid. The added struts were placed at 18" centers

Above: *mixing the mud.* Below: *Lamsin "dances" the circle.* At left: *the first circle, with sacrificial pot and feathers in foreground.*

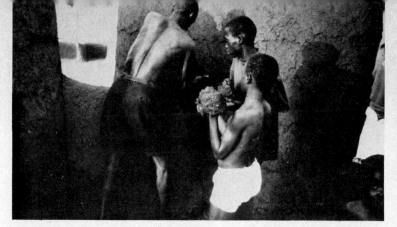

Above: *Lamsin working.* Below: *forming hen coops at goat house.*

around the top circumference of the earth drum. A rope was laid in a spiral over the rafters and lashed down to form a purlin support for a covering of thatching mats four inches deep. These were rolled on, tied back to the rafters and finished at the top with a woven grass knot or a broken pot. A good roof lasts for about three years.

Flat roofs were made with a basic structure of six inch diameter joists supported on forked timber columns equally spaced around the inside of the earth drum .... The columns carried the roof weight independently of the walls. A layer of two-inch joists placed over the base joists in the opposite direction were topped with twigs and bark; covered with laterite mud and a crust of laterite gravel. Laterite also makes immaculately smooth floor surfaces. Gargoyles to drain rain water were let into a parapet wall rising about a foot above the rooftop.

The first two rooms built had small shield walls attached to them. The first food was cooked here by Lamsin's wife. This T.Z. (millet gruel) is called dangerous food (sage bee). It was eaten in private by the cook and her husband. The food symbolised that he was now a man who could look after himself and guide his family through all life's hardships. Once the first two rooms were standing, the goats, sheep and cattle were confined well away from the area. If they had strayed onto the site and slept there before Lamsin and his wife, he would have broken down the construction and rebuilt it next year.

The house was fortified against such animal influences when the first room was cut open and Lamsin and his wife spent the night there together.

When the compound was completed Lamsin held a feast for all who helped raise it. He directed the Pito be brewed and himself slaughtered several guinea fowl. Young and old came together in the cool of evening-time and sat down to a special T.Z. containing meat (which is rarely eaten). The party soon became loud and joyful — children lithely danced and chased one another, lovers looked on, decked out in their brightest print clothes and old boys grated their cola-nuts on perforated sardine cans. The ancient Elder stood up. Calming the raucous enjoyment he turned his face to the failing light in the sky. He called on God to be present and grant riches and long life to Lamsin's family and all those who had toiled in the earth to make the new house.

On the day David and I parted we visited Bolgatanga market together, where he helped me buy a local smock. He told me that he studied at Agricultural college and expressed a desire to travel abroad and learn poultry farming to help his people — I believe he will probably do this. Perhaps soon, irrigation projects will bring water to this area making cash crops a possibility and subsistence farming culture will recede into the past. Hopefully a few 'traditionally' educated young men like David, who are not afraid of their culture, will help humanise 'modern progress.'

Farming

Farming, along with other human activity, is totally regulated by the climate. Dry season hoeing and weeding begin in early February and manuring (with highly prized animal droppings) carried on through the hottest months of March and April. Then the rains break in mid-May, with torrential burst and sporadic electric storms, violently cooling the unbearable humidity preceding them. The Volta basin becomes waterlogged five hundred feet below the well-drained fertile plateau farmed by the Nabdam. Here the Savannah woodland landscape of wizened trees uniformly dotted on monotonous plains of stunted, bleached grass tufts, explodes into lush greenery. Grasses that will eventually grow shoulder high are dotted with the colors of flowering trees. Early millet is now sown, followed by other high protein cereals (such as guinea corn, sorghum and maize) which are interplanted with various beans and Fra Fra potatoes.

Other crops include okra, groundnuts, sweet potatoes, gourds, melon, tomatoes and hibiscus. Further weeding continues through June, July and August; harvest time follows, with the cessation of the rains, from late September to mid-November ... every hour from sunrise to sunset is devoted to farming. After harvest the pace of life relaxes. Some of the young men go south to work in industries such as timber and cocoa; the remainder spend their leisure time dancing at harvest festivals or hunting with the old men. □

Rendille

The Rendille are nomadic shepherds of northern Kenya who travel with sheep and goats and carry their belongings on camels. In 1975, architect Anders Grum and his family lived and moved with the Rendille for seven months, photographing and documenting their life, "... in particular from a mobility and shelter point of view." These photos show the Rendille arriving at a new camp, unloading, and setting up their portable shelters.

Side elevation

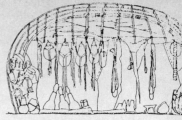

Longitudinal section

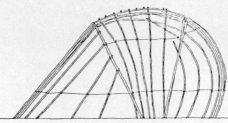

Cross section structure

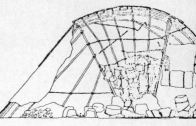

Cross section

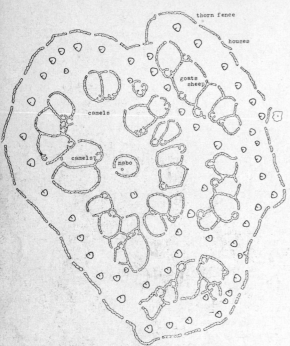

Plan of Gob Wambile - nomadic settlement of the Rendille

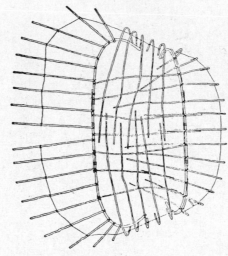

Plan of structure

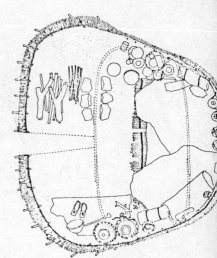

Floor plan

11

East
Africa

by Jack Fulton

Top left: *granaries for corn and vegetables in family compound in Lambwe Valley, Kenya. At left: home in village, Mwsambweni, Kenya. Below: structures of El Molo tribe, Lake Kenya.*

Above left: *Mwasam-bweni, Kenya.* Above right: *Lambwe Valley, Kenya.* At right: *school building under construction, Kwale, Kenya.*

Disaster Housing

by Ian Davis

Whenever a natural disaster takes place (and it is a frequent occurrence since one takes place every 85 days somewhere in the world), the media refer to the problems facing "homeless" victims. Then, in a major disaster, the public responds with their gifts to relief agencies and within a matter of hours jet transports are en route for the disaster scene. Inevitably there will be tents and possibly other forms of "temporary shelters" on board along with foodstuffs and medical supplies.

Recent research that has been undertaken for the United Nations Disaster Relief Co-ordinator (UNDRO) has revealed that there are rarely, if ever, "homeless victims." Despite the fact that their homes may be on the ground following some earthquake, or washed away in a flood, the victims of natural disasters show high levels of initiative in rapidly providing shelter for their families. This may involve moving in with friends or relatives, improvising a simple shelter out of anything at hand (probably rubble from ruined buildings) or the rapid reconstruction of their shattered home. Or it could involve all three activities.

Above: *improvised housing in Peru following the earthquake of May 1970.*

At right: *improvised shelter at a refugee camp near Calcutta following the mass migration of refugees from Pakistan in 1970. These sewer pipes provided temporary shelter for 12,000 refugees.*

At right: *the Amdanga refugee camp was set up in a temple courtyard in India during the Bangladesh disaster. 30,000 refugees stayed here in simple A-frames with thatch covering.*

At left: *improvised housing in Sudan following migration to search for food during the drought of 1973-75.*

Below: *a yurt-type shelter in the Republic of Djibouti, Africa in 1977. These refugees were displaced in the war in Ethiopia.*

House reconstruction six days after the 1976 Guatemala earthquake.

In February 1976, within 24 hours of the Guatemala earthquake, it is estimated that 50,000 of these shelters had been erected in the city streets and parks. Their form varied, some were made from rubble, others from normal city refuse—packing cases, scrap metal, etc. Their functions were to provide shelter for a family, a staging point for future action, a place for animals and surviving possessions. In November 1976 Turkey suffered its worst winter earthquake for 40 years. There were fears that the surviving population would die from exposure in sub-zero temperatures. However, despite considerable hardship there is no evidence of deaths from exposure. Some families interleaved several tents inside each other, thus providing a cellular protection from cold. Other families dug into the ground and put simple roofs over their holes, thus obtaining warmth from the soil below frost levels.

At left: *a relief parachute over an abandoned bus to form a temporary home in Dhekelia, Cyprus, 1975.*

Iquitos, Peru. When the river Amazon floods, which is an annual or biennial event, these houses built on wooden rafts ("Noah's Arks") float to the surface. Right: dry season; *below:* during flood.

These basic improvisational skills are a normal part of life within the Third World. In some instances these skills are used to prevent disasters from destroying homes, such as the "Noah's Arks" of Iquitos, Peru. Periodically the river floods and up float the homes!

These skills are crucial factors in recovery after a disaster. Western aid givers can learn to avoid placing obstacles in the way of these innovative skills. Such obstacles include: forcing survivors of a disaster to move to a new location away from their ruined homes and belongings; the burning or bulldozing of rubble (on the false assumption of disease risk) — thus destroying the essential building materials that families need for rebuilding; and finally free handouts — thus destroying local initiative which is a highly therapeutic process following the trauma of a disaster. □

New Guinea

by Ajit Mangar

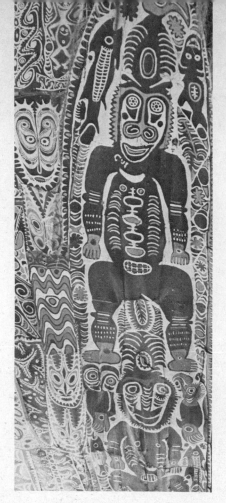

Located between the Australian and Asian continents, Papua New Guinea occupies the Eastern part of the island of New Guinea and the surrounding smaller islands to the East .... The individual family is the basic unit for gardening and care of children, but kinship groupings still play a very important role in the daily lives of the people. The tribal laws, public opinions, mutual obligations, and religious beliefs control the overall behaviour of the society.

Except near the town centres, houses are built of local timber, grass or palm leaves. The design and the construction techniques vary from place to place; some dwellings are round, some rectangular, some are built on stilts and some on the ground, which is used as the floor. The structures are light, earthquake resistant, climatically cool, and the walls and openings are ideal for ventilation, especially in the coastal areas.

Most structures decay due to weather and insects. The fire hazard is also great and large scale destruction of villages is often recorded.

The attitude towards houses varies greatly in different parts of the country. Some remain temporary huts while others are elaborate and skilfully constructed. In some areas no attention is given to the decoration, whereas in others it is considered as essential, and the artists and craftsmen are the most respected people in the community ....

Papua New Guinea is a nation of separate villages rather than a closely knit community with a national consciousness such as might be expected in any Western nation. The fragmentation exists because of the difficult topography, a great diversity of physical types, a total of 700 interrelated languages, self-sufficient traditional village economy, and warfare between the neighbouring tribes, which keeps people separate. Presently the society is living through a period of very rapid social change. For some groups, who have been exposed to the Western way of life for almost one hundred years, this has meant a drastic change in their way of life and attitudes. In other cases, because contact has been recent, very little has changed and the people live very much as their forefathers did.

Self government was achieved just over a year before independence in September 1975. All current development policy is guided and inspired by a set of aims approved by the Government in March 1973, popularly known as the eight point plan. The aims highlight the urgent need for decentralization, equal distribution of economic benefit, rural improvement and self-reliance. Within the context of this policy an urban housing programme was launched which covers a broad range of social, economic and geographical conditions. □

Floating Islands on Lake Titicaca

Urus

by James P. Warfield

The Urus Indians live some 12,500 feet above sea level on Lake Titicaca near the town of Puno, Peru. The Urus rely upon the use of one building material, *totora*, a reed which grows in the shallow areas of the lake. The Indians of this entire region of the high plains use totora to build *balsas,* reed boats made from bundles of reeds bound together. The reeds are often also used for sails. But the Urus are unique in that, having chosen to live on the lake, they utilize totora to build immense islands, which initially float on the water's surface.

Reeds are cut near the shoreline and towed in large piles to the island site. Eventually, a spongy surface is formed, capable of supporting weight. Shelter is built, also utilizing reed, and in time, earth is hauled from the mainland to create "fields" on the islands so that potatoes and other crops may be cultivated. Maintenance of the islands is a continual community effort, for as the reeds become water-logged, they sink. Thus, surface reeds must be renewed in order to maintain the island above water level. □

Pacific Islands

by D. Stafford Woolard

Despite limited resources and a low level of technology, traditional houses are often more efficient resource converters and climatic filters than modern structures. In many places, including the Pacific Islands, these techniques and skills exist today.

Can any developing country afford to ignore this great potential for housing solutions? As general energy sources decrease and the cost of imports from the developed world increase, the potential of the total local resources in housing will assume more importance.

On these pages are photos of housing in the Solomon Islands, the Fiji Islands and the Gilbert Islands.

Solomon Islands

The Solomons consist of high islands and low atolls. The climate is hot and humid, with regular winds and rainfall. There are two main housing types: the bush house, with low walls of vertical split timber, is found inland and in high areas; the coastal house, the most common, has woven wall panels, a verandah, and raised floor.

The house frame is constructed of timber culled from the bush. Wall and roof elements as well as the house plan are laid out on a one fathom module measured with outstretched arms. The house is generally built using prefabricated elements in about ten days as a community project. A conventional structural frame is built, then sheathed.

The coastal house has a raised floor built of split timber, laid round side up; thus sand and dirt fall through, and breezes are drawn through the floor for cooling.

Walls are lashed to the framework with vine and in many cases are designed to blow off in cyclones, thus protecting the main house from destruction.

The beautifully detailed roofs are made of ivory nut leaf, folded over and stitched along a bamboo rib. The panels are dried and lashed to the structure; the closeness of the panels determines the quality of the roof, the distances being measured with the fingers. Many roofs are replaced in ten years, but a well-built roof can last 25 years.

Fiji Islands

Fiji's climate is temperate, with high temperatures and consistent trade winds. House forms relate to the three main island climates. In the coastal wet zones, with high rainfall and humidity and small diurnal temperature range, houses are lightweight, well ventilated, with well insulated roofs. In the coastal dry areas, with continuous trade winds, average rainfall and humidity, houses have more insulation in the walls. In the highlands, with varying rainfall and large temperature variations, the houses have thick walls and roofs, with little ventilation. The construction of a traditional Fijian house is undertaken as a concentrated effort involving many people over a short period of time. Once completed, the structure is sealed and a fire lit inside which forces smoke into all parts of the thatching as a deterrent for bugs and beetles. The main structural frame of these houses is heavy logs, with bamboo used as secondary material, resting on a stone base. The earth floors are built up above the surrounding ground with stone retaining walls, and covered with grass mats. The walls in early houses were very low, sometimes to the ground. In coastal areas walls are woven split bamboo to allow breezes to penetrate. Other wall materials are woven reed or grass, and reeds and bamboo compacted into a dense material. The steep roof pitch not only sheds water well, but allows only the ends of the thatch to be exposed, thus increasing the roof's durability. With regular maintenance and smoking, the roof will last 20 years. An important part of the Fijian roof is the free form trunk, which protrudes from each end of the ridge.

Pacific Island structures place most emphasis on the roof. Most time, effort and thought is given to that part of the house that excludes the climatic extremes of sun and rain.

Gilbert Islands

Located on the equator, the Gilbert Islands are comprised of 16 atolls, which have a maximum height of five meters above sea level. The climate is warm and humid with frequent trade winds blowing. Rainfall is seasonal with long periods of drought.

Traditional house forms are consistent throughout the islands; the basic house consists of three parts: sleeping, storage and cooking areas. The most important part is the sleeping area, which has a raised floor, minimal walls, and low roof. There is no internal subdivision of space and no internal furniture or fittings.

The traditional Gilbertese house is largely prefabricated, with specialists completing each component. The pieces are then assembled on site quickly. The main structure is of pandanus timber with subsidiary members of split coconut and coconut mid rib. Walls are usually woven panels hung on strings from the roof structure. The most substantial part of the house is the roof structure which is prefabricated on the ground, then lifted on to the four corner posts. Roof pitch is about 45° and there are small ventilating areas built into the ridge. The roof form is hip ended.

Conclusion

In all cases described the houses have been constructed from renewable local resources which were available at no cost. All houses seem particularly well suited to the climate, the living patterns of the people and skills developed through time.

In fact traditional houses are often more advanced in technical concepts and thermal comfort than the designs which are replacing them. These modular houses are constructional systems and are independent of the house plans. As people become urbanised and desire an urban subdivided form of lockable dwelling, these would be quite feasible using traditional techniques. □

*Sloping gutters on wall channel water
into underground storage cisterns.*

Menorca

Photos by Regan Bice and Josep Mascaró

On Menorca, one of the Balearic Islands off the coast of Spain, is a unique collection of old farmhouses, built in phases over long periods of time. Since wood was scarce on the island, building materials were rubble rock, quarried limestone, ceramic tiles for roofs and floors, and limited amounts of wood for ceiling beams, door, and windows. Exterior and interior walls and ceilings are whitewashed annually to protect the masonry.

Due to a shortage of water, rain water is collected from the roofs and funneled into underground cisterns. Sloping gutters are built into the walls, or inverted roofing tiles are mortared together, circling the entire house to direct water into the cisterns. Farmers often block the first rain of the season, so that the roofs and gutters are cleaned before water is collected. The cisterns are often small quarries, having supplied the limestone for building the house. They hold up to several thousand liters, enough to last a family through the dry summer until the next rainy season. □

Turkish Yurts
by Suha Ozkan

There are two main types of mobile shelters of the nomadic Turkish communities. One is the black tent (*kara-kadir*) a tension structure of goat hair fabric and ropes supported by three or more posts. These tents can be seen almost anywhere in Anatolia. The other type structure, now rare and limited in number is the dome shaped tents with ribbed wooden framework and felt covering: known in the Western world as *yurts,* and in Anatolia today called *alaciks.*

Throughout history the alacik has been utilized extensively by both Turks and Mongols, with the general consensus being that the shelter is of Turkish origin. Different Turkish communities give it the same name with some differences in pronunciation. In Anatolian vernacular the word alacik is sometimes used as a general reference to shelter.

Archaeological traces of alaciks go back to the Huns of Asia. And one can even today observe yurt type shelters almost identical to those in the 16th century miniatures and battle scenes.

The alacik was adapted to Central Asian climatic hardships. The need to search for fresh pasture required mobility, the extremes of hot and cold called for alternate insulation and ventilation, and strong winds of the steppes necessitated structural stability. The ribbed skeleton and domed structure with multiple layers of felt and varying venti-lation possibilities, makes the alacik ideal for this climate, and for nomadic existence. Alaciks are the mobile shelter type in the Turkish states in the Soviet Union like Uzbekistan, Turkestan, Kir-ghizistan; in Sin-Kiang in China; in Northern Afghanistan, Iran, and Anatolia.

There are two basic types of alaciks:

— *catma ev* (frame house): the primitive version, with elliptical or circular rib-bed-cage skeletal structure. Its wooden sticks are curved into U-shapes, and both ends planted firmly in the ground to form the basic structure.

— *topak ev* (round house): miniaturized version of Central Asian yurts. The frame of one type consists of 20 vertical and 20 horizontal ribs of flexible juniper wood, the ends sharpened for driving into the ground. Horizontal ribs are tied to verticals with hair rope or leather thongs, making a rigid lightweight structure. This is covered with several layers of felt, then tied around with ropes for the winds. To prevent wind uplift of the felts, stones are hung from a skirt attached to the circumferential ropes.

The nomadic *topak ev* dwellers do not carry the alacik ribs with them when travelling: each family has two sets of sticks. One is hidden in the highlands, the other in the Mediterranean lowlands. This eliminates the major burden of moving.

The north side of an alacik is generally the storage area, with the entrance opposite, to the south. The material stored to the north provides insulation on the coldest side. Milk or dairy products are stored close to the entrance, away from the fire. There is usually an opening at the apex of the alacik, opened and closed alternately for ventilation or heat retention.

The more advanced *topak ev* type alaciks are now almost extinct in Anatolia. They are used by Turks who immigrated from Central Asia, as summer dwellings while grazing.

In structure, the *catma ev* is a doubly-curved framework, while the *topak ev* is more like an umbrella.

Central Asian weather changes are adapted to by having a single felt layer in the summer, five to eight layers in the winter. In today's surviving alaciks polyethylene is sometimes used for weatherproofing, with inner layers of felt.

An alacik *with a circular plan in Korkuteli, Anatolia* (Catma Ev).

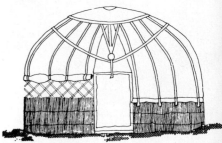

Turkmen (Afyon, Anatolia)

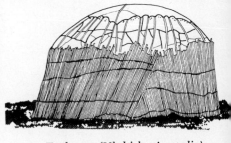

Turkmen (Ulukisla, Anatolia)

…omadic community of alaciks *camped on Bey Mountains, Anatolia (Catma Ev* type*). Stone shelters built for livestock are permanent.*

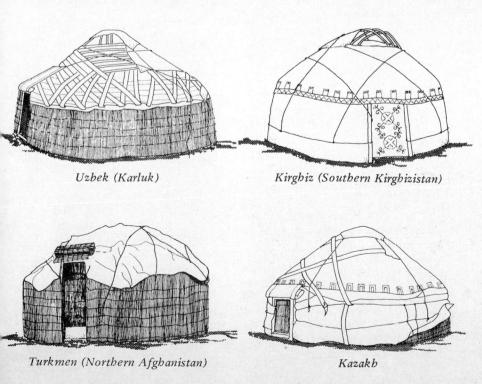

Uzbek (Karluk)

Kirghiz (Southern Kirghizistan)

Turkmen (Northern Afghanistan)

Kazakh

Erection of topak ev type alacik:
- stretching the *keregu* grills.
- placing *keregu* to form cylindrical base.
- erecting center post.
- putting *dugnuk* (roof structure) on top of post.
- inserting *ug* sticks to *keregu*.
- tensioning wooden skeleton by tying with horizontal and diagonal belts.
- covering with layers of felt.
- placing top leather piece with radiating belts.
- connecting these belts to bottom circumferential rope.

This process takes about 25-45 minutes. When the shelter is moved it never exceeds one camel or mule load.

For summer ventilation the layers of felt are rolled up and tied so the sides are open, or they are maneuvered so there is cross ventilation between the sides and the top opening. Reed mats *(ciq)* are usually wrapped around the open alacik framework in summer as insect barriers while a steady flow of air is maintained.

The wooden skeleton of the *topak ev* lasts about 20-25 years while the felting must be renewed every 6-7 years.

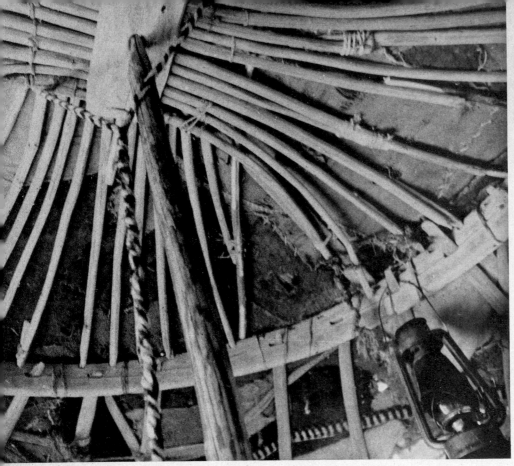

Topak ev *type* alaciks *are prestressed structures: vertical tensioning is provided by a central post supporting the weight of the roof* (dognuk) *and a rope tie-down for wind uplift. Flat curved sticks radiating from center are* ug *sticks.*

Connection of ug *sticks to* kerege.

Turkmen children

The family structure of alacik dwellers differs from that of the Eastern Anatolian nomads who live under "multi-poled black tents". Whereas the tent-dwellers live in extended families, there are independent nuclear families living in alaciks. In the tents a whole tribe will occupy only a few shelters; livestock are kept in the same tents but in different cells. The dominating and old people occupy the centermost cells, surrounded by the younger members. The tents are divided by vertical hangings. The arrangement of animals, young and old has these advantages:
- the elders and rulers are farthest away from invaders.
- the successive concentric chambers form heat barriers so the middle is warmest.

By contrast, alacik dwellers do not share interior space with their animals, but sometimes build them covered enclosures.

Socially each alacik symbolizes a family. After marriage, a new alacik is built for the new family. With live-stock donations from both bride and bridegroom's families a new economic and social unit is formed. The independence of each family is highly respected, yet this does not preclude communal life. Families join for cultural and social activities, and move together. A community of independent families, each a sufficient production unit, means a community with no landlords or oppressive order. Problems are solved at the family level.

Centuries of use of this type shelter has made them acquire symbolic content and value. The spherical form signifies the sky. The old Turks used to believe the column of the sky was Polaris (*universalis columnu* in Greek and Roman cultures). The Turks' concept of the universe was similar to that hypothesized by Ptolemy, except they conceived of Polaris in the center, instead of the earth. At the time when their religious belief was Shamanism, the shaman's climb up the center post to reach occulus symbolized a climb to the sky to reach heaven. □

Mongolian Yurts

Many changes have taken place in the lives of nomadic Mongols since the socialist revolution of 1921. One area of life where changes can be easily seen is in the family dwelling, the round felt tent, called a *ger*, which is still used by most of the population. A young Moscow-trained Mongolian ethnographer, G. Tserenxand, recently charted these alterations over the past couple of generations in central Mongolia.

Until recently, the family was not only the main unit of ownership and production in herding, but it also organized its life in an exceptionally rigid and formal manner, closely tied to the old social conditions. Categories of age, sex, genealogical seniority, wealth, and religious status were maintained by explicit rules and prohibitions within the domestic circle. The round tent was virtually the only dwelling known in Mongolia, apart from Buddhist monasteries, and it was the focus for relationships between people widely separated by daily occupations. It provided a space in which every category of person or object in the nomad's world could be located, and so became a kind of microcosm of the social world of the Mongols.

The space within the tent was basically divided into four sections, with the fireplace being the cut-off point. Half of the *ger*, from the door (facing south) to the fireplace, was the area of lower status; that behind the fireplace was the more honored half. These halves were then further broken down by sex: the male, or ritually pure area, to the left of the door, and the female, or unclean quadrant, to the right of the door. This spatial sectioning, in turn, determined the placement of the various household furniture etc. around the perimeter of the *ger*.

In spite of the fundamental changes that have occurred since the revolution, people still live in family groups in felt tents.

Traditional *ger*
Clockwise around perimeter, from door: saddle; bridle and halter hanging on post; sour mare's milk hanging in leather bag; area for storing felts, skins, blankets, bought food; chest belonging to master of household; gun or other weapons; Mongolian and Tibetan books; Buddhist altar on chest which contained money and other precious items; wife's chest, hat box; marital bed (young children might be at lower end in a pen); eating utensils, cooking pot.

In center: brazier, surrounded by felt mats and skins; low wooden table for serving tea and food.

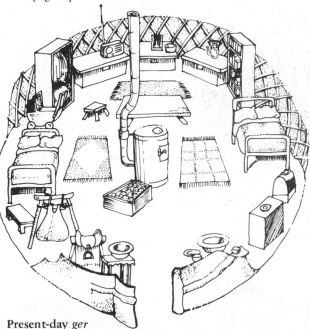

Present-day *ger*
Clockwise around perimeter, from door: washstand; saddle; sour mare's milk hanging in leather bag; children's bed; baby carriage; wardrobe; chest for master of house with radio on top; chest for valuables; chest for women's clothes; bookcase; marital bed; portable sewing machine; cupboard with china crockery; cooking utensils.

In center: iron stove with door facing east (towards wife's place); wool rugs; low table for tea and food; stool for guests.

Adapted from an article in *New Society* magazine, October 31, 1974, by Caroline Humphrey; drawings by John Storey.

Syria

Syrian nomads, who travel with flocks of sheep and goats, call their tents *Beit Charr*, translated: *House of Hair*. They weave either goats' hair or sheep's wool into unit webs about two feet wide and the length of the tent. Ten or so of these webs make up the tent, which is usually of two colors: black and brown. They are laced together, reinforced by narrow tension bands, propped up by poles and staked to the ground.

The tent walls are stitched at the sides of the main tent with steel pins. On the leeside of prevailing winds it is left open during the day; thus the enclosure is protected from wind and dust by walls on three sides. These are usually one-family tents, divided internally into two areas, men's and women's, by setting up reed mats; each of these areas contains a fire pit — one for coffee making, the other for cooking. Sleeping is provided by laying mattresses on the carpets which cover the ground.

Daytime use shows the unique character of a Bedouin's concept of living space. With the entire length of one wall usually open, the people enjoy the shaded semi-open space, and the exterior area becomes a part of the living and working space. □

Tents

by Koji Yagi

Turkey

Syria

Mediterranean Sea

Turkey

Three-poled black tents in Bey Mountains, the highlands of Antalya, Turkey.

Syria

Greece

by Aris Konstantinidis

"...there is only one starting-point: the soil, the stones, the water and air of Greece. From there you must begin...."
Kazantzakis

And gradually, as each new coat of white-wash covers the previous one, these surfaces seem to acquire a soft, warm, human-like skin, you can lean against it and touch it and feel it breathing.

The Greek light, that gives such a brilliant aspect to every constructed surface. The Greek light, "that transparent, weightless light, full of spirit, that clothes and unclothes all things."
Kazantzakis

In summer, when the sun is high, the open roofed area or colonnade stops the rays from penetrating into the closed inner space or rooms, and helps to retain a measure of coolness inside the house, as well as to keep out the blinding light. Conversely, in winter, when the sun is low, its rays slip through the open-roofed area or colonnade into the inner space, thus bringing warmth and abundant light; provided of course that the roofed area faces south, for in Greece this is the most favourable position for any type of living quarters . . . .

. . . in a country like Greece (and in other countries with similar climates), people indeed spend most of their time (both leisure and working time) under colonnades or covered verandahs, i.e., in semi-outdoor, intermediary areas . . . .

Consider the beauty of this ceiling. . . The insulation of flat-roofed structures from heat and cold, which modern builders achieve by means of artificial materials, was done equally well by builders of the past with natural materials. The ceiling's beauty has much to do with the linear patterns of the roof-beams, and that 'breathing quality' which we find again, strangely enough, the pattern made by the steps of a village street in another island. How true, then, that no matter what material a civilized craftsman (in the true sense of the word) takes into his hands, he will consistently produce shapes and forms expressing a similar spirit, possessing the same simple and noble quality.

In the Greek landscape of today, we often come across buildings that are modelled with amazing closeness on the monuments of ancient Greece; as if they had been erected on the soil thousands of years ago . . . . The dovecots of Mykonos . . . now stand forgotten and derelict, as if their form were slipping away irretrievably, now that they have apparently outlived their function.

38

These threshing-floors, spread across the mountain slopes — are they not reminiscent of ancient theaters?. . .

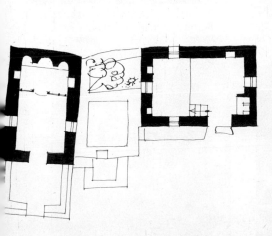

Photographs, drawings and notes from *Elements for Self-knowledge* by Aris Konstantinidis (*see bibliography*). □

CARAMAGNA P.
TOWN PLAN

3 EXPANSION
NUCLEI c.XVI–XVIII

ORIGINAL
NUCLEUS
(FOUNDED
AFTER 1028)
FULL CHARTER
GRANTED 1435

N

Plains Towns of Northwest Italy

by John Hamilton Doyle

The hill towns of Italy are so renowned they quite unjustly overshadow the many plains towns in the Po valley and elsewhere. These are just as remarkable, if less picturesque (being more difficult to appreciate from the ground), and although the peasant society and culture which created and sustained them for over 700 years has all but disappeared, they still offer valuable lessons today. They can be likened in several ways to a modern urban squatter settlement, especially the medieval new towns of 1200 to 1350 A.D.

1. The economy of scarcity, or lim-

Above: *typical north-south street.*
Below: *typical east-west street.*

Aerial view of Caramagna looking north.

ited surplus, made it imperative to be economical with land and materials. This was a fundamental principle behind the form of these settlements.

2. To make a new settlement a success it had to quickly become a viable community. To attract settlers, the founders offered benefits such as free land, a town charter, citizens rights, etc. These plus the "negative" stimulus of outside threats to safety (like police opposition to squatter settlements) all contributed to the community's cohesion and identity.

3. Any internal dissent which would have risked the new town's success had to be avoided. This was important in the distribution of land, which had to be done fairly. This, plus the primitive surveying tools available, dictated a regular orthogonal layout; employing circles and the like would have led to intolerable levels of approximation.

4. Later expansion of the town also followed this type layout because the fields around the original nucleus were organized in the same way. The actual perimeter of the town varied frequently in shape — imposed by topographical features of the site. No two towns are identical.

5. Land was distributed according to the needs and size of each founding family. Within the town each lot was cut from the long E-W oriented blocks typically 25-30 metres deep which ran off the main N-S street like the teeth of a comb. As population increased the original nucleus, hemmed in by its walls, grew more and more crowded and today the original clarity is often lost, modified also by changes in function.

6. How economical and sensible these illiterate peasants often were is shown by the way they built and placed their houses in rows (see sketch).

7. These people built their own housing, just as squatters do now and just as millions will continue to do; their own solution to the housing problem! Starting with a sort of "invasion shack" this was gradually converted into a permanent home.

8. The layout and housing, though not without problems, is generally valid for living today, accommodating our modern services (bathroom) and machines (cars) easily enough. Modern building codes and zoning regulations often seemed weighted against them (ceiling height minimums, etc.) and people now want individualistic and isolated villas, so the tendency is towards their abandonment, which is a pity, considering their efficiency and good sense. They still serve as a lesson today. □

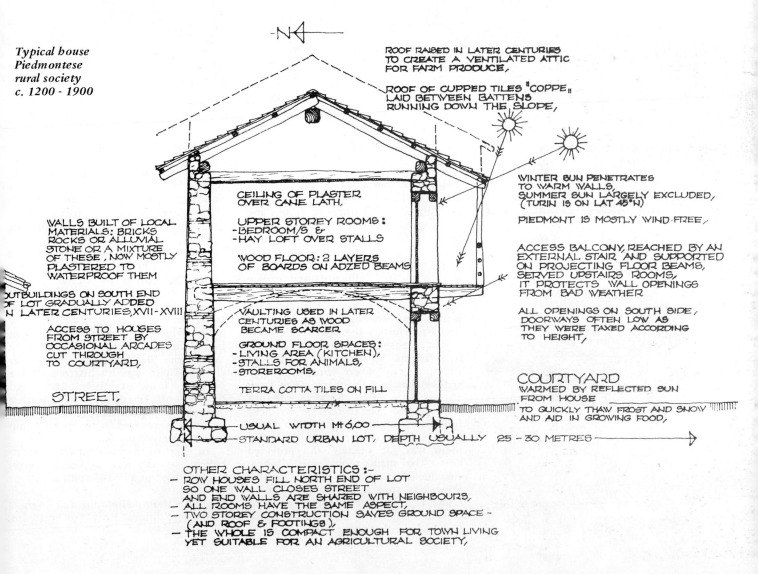

Typical house Piedmontese rural society c. 1200 - 1900

-N↑

ROOF RAISED IN LATER CENTURIES TO CREATE A VENTILATED ATTIC FOR FARM PRODUCE,

ROOF OF CUPPED TILES "COPPE" LAID BETWEEN BATTENS RUNNING DOWN THE SLOPE,

WINTER SUN PENETRATES TO WARM WALLS, SUMMER SUN LARGELY EXCLUDED, (TURIN IS ON LAT 45°N)

PIEDMONT IS MOSTLY WIND-FREE,

ACCESS BALCONY, REACHED BY AN EXTERNAL STAIR AND SUPPORTED ON PROJECTING FLOOR BEAMS, SERVED UPSTAIRS ROOMS, IT PROTECTS WALL OPENINGS FROM BAD WEATHER

ALL OPENINGS ON SOUTH SIDE, DOORWAYS OFTEN LOW AS THEY WERE TAXED ACCORDING TO HEIGHT,

CEILING OF PLASTER OVER CANE LATH,

UPPER STOREY ROOMS: -BEDROOM/S & -HAY LOFT OVER STALLS

WOOD FLOOR: 2 LAYERS OF BOARDS ON ADZED BEAMS

WALLS BUILT OF LOCAL MATERIALS: BRICKS ROCKS OR ALLUVIAL STONE OR A MIXTURE OF THESE, NOW MOSTLY PLASTERED TO WATERPROOF THEM

OUTBUILDINGS ON SOUTH END OF LOT GRADUALLY ADDED IN LATER CENTURIES, XVII-XVIII

ACCESS TO HOUSES FROM STREET BY OCCASIONAL ARCADES CUT THROUGH TO COURTYARD,

VAULTING USED IN LATER CENTURIES AS WOOD BECAME SCARCER

GROUND FLOOR SPACES: -LIVING AREA (KITCHEN), -STALLS FOR ANIMALS, -STOREROOMS,

TERRA COTTA TILES ON FILL

COURTYARD WARMED BY REFLECTED SUN FROM HOUSE

TO QUICKLY THAW FROST AND SNOW AND AID IN GROWING FOOD,

STREET,

USUAL WIDTH Mt 6,00 — STANDARD URBAN LOT, DEPTH USUALLY 25 - 30 METRES

OTHER CHARACTERISTICS :- - ROW HOUSES FILL NORTH END OF LOT SO ONE WALL CLOSES STREET AND END WALLS ARE SHARED WITH NEIGHBOURS, - ALL ROOMS HAVE THE SAME ASPECT, - TWO STOREY CONSTRUCTION SAVES GROUND SPACE - (AND ROOF & FOOTINGS), - THE WHOLE IS COMPACT ENOUGH FOR TOWN LIVING YET SUITABLE FOR AN AGRICULTURAL SOCIETY,

Celtic Dwelling

by M. Pierre Gac

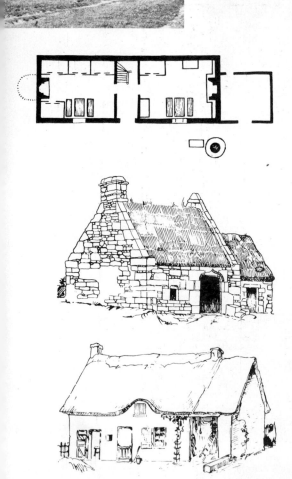

The people of Brittany, descended from the Celtic civilization, developed a type of dwelling (called *ti* in Breton language) which was found in various parts of Brittany. This farm dwelling was built to resist severe weather conditions and for centuries has been the House of Life for people living in harmony with their natural environment, cattle and poultry. The Breton farmers were not affected by any foreign influences and had achieved a certain serenity through the natural rhythms of their lives, and hard work. Their dwellings were scattered in detached clusters throughout the land.

The *ti* was composed of two ground-floor rooms, separated by a corridor and a granary where corn, vegetables and appliances were stored. The floor was of bare earth. Since hospitality was a sacred rule there were no locks for the doors. The room shown above was the heart of the house; the other room was for animals who provided heat. The main room was kitchen, chamber, parlor and pantry. Everything was stowed in traditional positions. The fireplace was the center of everyday life, and used for cooking (pots were either hung from a hook or set on a tripod), heat and often the only light. It was the center of attraction at the winter evening meetings, when the grandfather sat on a bench under the chimney and told tales to the children while the parents were preparing the candles or spinning the hemp.

The common room was partitioned by the box-bed, which had sliding doors that could be closed. It was built this way because of the danger of pigs or chickens (which roamed freely in the room) injuring or killing babies while the parents were at work in the fields. A cradle could be hung inside the box-bed and the doors closed. There was no mattress, but a *paillasse* made of broom, straw, reed or sea-weed. An oat-chaff filled feather bed was laid upon the *paillasse* and the whole was covered with rough hemp sheets and green blankets.

In front of the box-bed was a chest-bench, used as a seat, for bread storage, a step ladder, or a bed if needed.

From *La Maison Bretonne* and *Logis et Menages*, Editions d'Art, Jos Le Doare, Chateaulin, France.

Everyday life was punctuated by a series of fests and customs which corresponded to the collective work: the construction of a building, or the harvest were opportunities to kill a pig and drink a lot. Or the ground ceremony: when the mud floor became too uneven or muddy, people from the surrounding farms were invited on a Sunday. A mixture of clay, ashes and cow manure was spread upon the ground to harden it, and the dancing and stomping packed the floor for the next year. □

Ar C'havel.   Le Berceau   *The Cradle*

Les Maisons Paysannes Francaises

L'Art de Restaurer une Maison Paysanne par Roger Fischer (voir bibliographie) est un ouvrage impressionnant au sujet des maisons de campagne. Le livre commence par un avertissement qui plaid pour une restauration sensible, et non pas une fausse restauration dite rustique, qui détruit le véritable caractère de ces maisons.

Il y a plus de 500 excellentes photographies de bâtiments paysans, avec des exemples de toutes les régions du pays. Les éléments dont les maisons et leurs abords sont composés sont montres en detail; clôtures et barrières, murs et toitures faits de divers matériaux, fenêtres, portes et volets, balcons et galeries et intérieurs.

Ce n'est pas une livre à propos du métier de restaurer; il n'y a pas de détails de construction. Le sujet est l'intégrité et l'esprit des maisons paysannes, des exemples qui existent déjà, et les possibilités de préserver ce riche héritage culturel.

Ci-dessus: *escalier extérieur d'une chaumière du Morbihan.*
Ci-dessous: *il n'est plus temps que quelqu'un vienne sauver cette humble maisonnette: le bulldozer y est passé. Aux habituels caractères percherons, elle joignait la noblesse de sa toiture haute aux pans curieusement assemblés. La maison percheronne "pousse" volontiers dans tous les sens, par adjonctions de laiteries, bûchers ou étables, et tire souvent de cette imbrication de murs et de pans de toiture une véritable beauté architecturale.*

Ci-dessus: *baie intérieure ouverte entre deux pieces. Sa réussite est due à sa simplicité, et à l'emploi de pièces de bois provenant d'une charpente ancienne.*

Ci-dessus: *en Bresse, galerie extérieure "au sol," caractéristique de cette région.*
Ci-dessous: *rampants à "pas de moineaux" constituant de véritables marches d'escalier, couramment utilisées pour l'entretien du toit (Bigorre).*

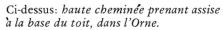

Ci-dessus: *haute cheminée prenant assise à la base du toit, dans l'Orne.*
A droit: *charme particulier des toits de laves imbriqués sur cette grosse ferme des Causses.*
Ci-dessous: *une triste maison paysanne d'aujourd'hui; elle garde tout son caractère et même son four à pain, mais est abandonnée au bétail et tombera bientôt en ruine.*

Ci-dessus: *petite porte donnant accès au grenier d'une maison de Corrèze.*
Ci-dessous: *rive grossière de chaume sur un mur de pisé (Bretagne).*

English Cottages

Drawings by Sydney R. Jones

There is probably no object so much a natural part of the English landscape, nor which makes such a direct appeal to the heart and imagination, as the old country cottage. In every part of England, in the village and on the outskirts of the town, in the hamlet and standing on the lonely moor, there still remain these beautiful witnesses to the vitality, freshness, and pride of the village mason and carpenter. Passing from district to district the wonder grows at the many types, and that half a day's journey from cottages of stone there are cottages of cob and thatch . . . .

The craftsman with more imagination than his fellows gave a new turn to the mouldings, finished a gable with a finial of a fresh pattern, or added another variety of walling; one carried through his work a little in advance, and one remained a little behind, but the work as a whole was customary and usual, and the following on of what their fathers had done before them. Each gave of his best, his quota of simple and direct workmanship, using the materials that were to hand, sometimes wisely and well, sometimes badly, but always inspired with a fancy and invention as natural as they were unconscious. The way they built and the way we build are essentially different. With them the tendency was to add gradually new methods of doing things, slowly increasing their store of ideas, from which they drew, as they drew water from the well on the village green . . . .

The plan seems to have had an origin quite distinct from that of the circular hut. At first it was merely a copy of the simple rectangular structures erected for the housing of the oxen. It was built in bays to accommodate what was called a long yoke of oxen, that is four abreast, and the bays divided by two pairs of bent trees, in form resembling the lancet-shaped arches of a Gothic church, and placed at 16-foot intervals . . . .

In the arrangement of materials, whether of one or of many, the village workman displayed a happy knack of doing the right thing in the right place, but in putting them together he was not always so successful, and seldom satisfactory from the sanitary expert's point of view. The rain was allowed to drop from the eaves without any means to collect it, the water to sink into the foundations, and walls were sometimes badly built; but in spite of these drawbacks, and possibly partly owing to them, the appeal of the country cottage is universal . . . . □

From Old English Country Cottages, *edited by Charles Holme; published by The Studio, London, 1906.*

Ireland

Gypsy Vans

by Denis E. Harvey

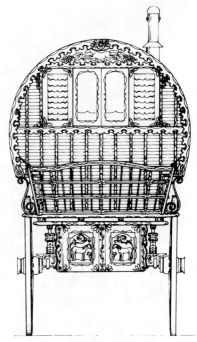

Denis Harvey is co-author and illustrator of The English Gypsy Caravan *(see bibliography). The book is the result of 25 years research by the authors. The Wright bow-top van described here is ". . . the most typical and highly evolved spin-off from the gypsy scene in England. . . . These vans were superbly designed and a few highly-prized survivors are still in use in the old style by horse-drawn travelling people. . . ."*

The domestic solution for Gypsies in Britain from the late nineteenth century onwards was this Bow-top Living-wagon designed and built by William Wright & Sons of Leeds, Yorkshire, England. Bill Wright's bow-tops became the type of van most favoured by horsedrawn travelling people other than Showmen due to its functional elegance, easy maintenance, and under-body clearance combined with a low centre of gravity.

Front

Rear

Bow-top Waggon

Wright c 1906

11 feet long
10 feet high

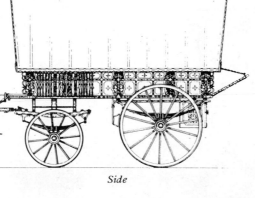

Side

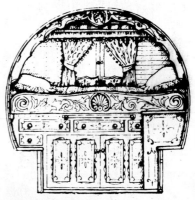

Interior facing bed

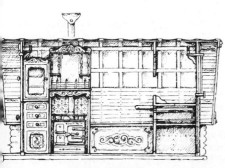

Interior left side

Cooking & Sleeping: These wagons would be used by Gypsy families in conjunction with bender-tents and 'accomodation' carts. Although they contained cast iron cook-stoves, cooking was mostly done on a stick fire outside with pot, kettle and skillet hung from kettle-props (see photo) — one of the most efficient and economical ways of cooking for those who become familiar with it. □

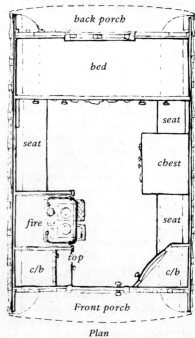

back porch

bed

seat

seat

chest

fire

seat

c/b   *top*   *c/b*

Front porch

Plan

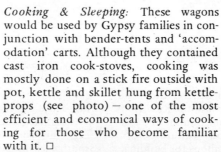

Interior right side

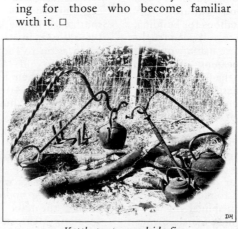

Kettle props, roadside fire

49

Kickapoo

by Peter Nabokov

Kickapoo Indians migrating to Mexico (date unknown).

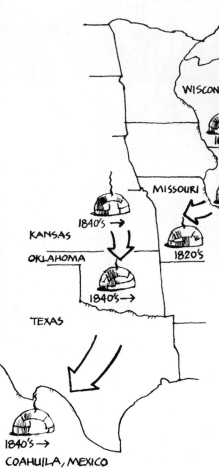

Dispersal of Kickapoo housebuilding tradition.

In the beginning, the Mexican Kickapoo believe, their all-creator, *Kitzihiat,* lived in the prototype *wikiup,* the Kickapoo winter house, and the prototype *odanikani,* summer house. He showed his son, *Wisaka,* the Kickapoo "culture hero," the materials and techniques, rituals and taboos for building, consecrating and living in these houses. In turn *Wisaka* taught them to the remarkable Indian tribe whose name would fit their peripatetic history: *Kiwigapawa,* meaning "He Moves About," simplified to Kickapoo. Wherever the Kickapoo wandered thereafter, the maintenance of these house-building customs and codes was essential to tribal integrity.

Throughout three centuries of migration down 1,500 mid-Western American miles from the Great Lakes to Mexico, the Kickapoo have preserved to this day their traditional house forms as the most vivid, visible badge of their fierce adherence to old ways. It is common for house forms to undergo drastic changes during times of cultural stress or relocation – not so with the Kickapoo. Despite adaptations to radically different climates, they have kept up their semi-annual change of dwelling pattern, their old customs regarding female owner-

ship and building houses, the consecration ceremonies and house-use customs.

Originally the Kickapoo lived west of Lake Michigan, but early distaste for white culture prompted them to migrate southward in the late 1600's. First they built their lodges in Illinois, next in Missouri and Kansas. Shortly after they moved south again to Indian territory, and some small bands meandered into old Mexico. In 1842 the Mexican government formalized their hands-off policy toward the Kickapoos by granting them 40 square miles south of the Texas border in the state of Coahuila. Other American Kickapoos in search of cultural freedom joined their Mexican kinsmen in 1863, until today the unlikely desert area boasts the last pocket of 19th century Woodland Indian lifestyle left in North America.

Today the Mexican Kickapoo still build their tule or cattail mat lodges – photographed in 1952 by two Wisconsin anthropologists who were allowed a rare, two-week sojourn before being booted out. Long a legend in anthropological circles, the tribe has, however, made a few accomodations to their Mexican surroundings. Here and there a "jacal"

Mexican Kickapoo settlement in Coahuila, Mexico. 1952.

(cornstalk or saguaro cactus Mexican rural house) joins the Winter house, Summer house, Cookhouse and Menstrual hut which comprise the shelter grouping. And the family cluster resembles the traditional Mexican "solar," or family compound.

Crossing the border at Eagle Pass, Texas, the Mexican Kickapoo enjoy a unique status. They are allowed through customs without green cards or tourist visas as they undertake yearly trips to visit stateside Kickapoo communities in Kansas and Oklahoma. Their reputation for hostility towards outsiders — particularly those bearing note pads, cameras, and tape recorders — is notorious, but a generous couple named Mr. and Mrs. Lewis Cuppawhe kindly invited us on a tour of their winter lodge and summer house frame. A hundred yards from an Interstate, Mrs. Cuppawhe demonstrated the sewing of cattail mats using a cow rib needle, and a mile from the Shawnee, Oklahoma supermarkets, she showed how she ground dried corn in a homemade oaken mortar.

continued

Mr. and Mrs. Lewis Cuppawhe. 1976

The typical Mexican Kickapoo compound encloses within its barbed wire fence five traditional structures: 1) the haystack-shaped winter *wikiup*, with its double-thick mats battened with outside stringers and laid so rain will run downward; 2) the peaked summer *odanikani* with its vertical walls—made from sotol flower stalks—and two mat layers, the one laid in an east-west position, the other north-south; 3) the shade, or *ramada*, with free-standing work benches, open on all sides and connected at the western eave to the summer house entrance; 4) a 12 foot-square cook house with tough framing of juniper (like the summer house), roofed with hard Beaked Yucca; 5) toward the compound rear, a menstrual hut where women live during the time of their period, about six foot square and containing bedding and implements used here exclusively.

Around the living space the ground is of tamped earth, frequently sprinkled. While the dwellings themselves are individually owned, the compound space is the property of the entire village; occupants have lifelong rights there so long as they observe the house-upkeep code.

Winter house (wikiup)
October to March occupancy
20' long x 14' wide x 9' high woven frame

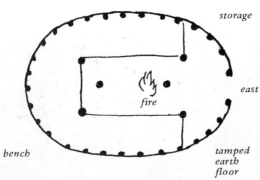

Interior Furnishings:
    Bedrolls: *Quilts and blankets on benches*
    Trunks: *for bedding storage during the day*
    Crude chairs
    Treadle sewing machines
    Flashlights, mirrors, hanging from inner frame
    Bench: *dyed and woven cattail mats on sapling frame raised off earth floor 4"*

The *wikiup* frame is made of hackberry, Montezuma bald cypress, or sycamore saplings bent over a rectangular support of four sycamore posts buried in the tamped-earth floor. The horizontal pieces are then tied to make a framing lattice-work — all joints tied with *pita*, shredded and wound yucca-like fibers.

The cattail or tule mats are first wrapped — with stalks up and down for rain run-off — around the base of the wikiup frame, then laid over the top to leave a six-foot long smoke-hole slit. Smoke-darkened and worn mats are put on the exterior while newer mats face the inside. Finally the mats are battened down with stringers.

52

Summer house (odanikani) *with ramada*
*March to October occupancy. 16 -18' long x 15' wide x 11'
high. Hipped roof on vertical walls.*

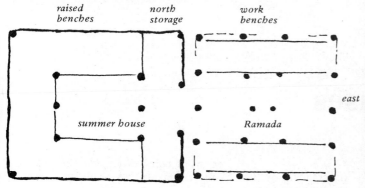

Interior furnishings:
*Workbenches for sewing, basketmaking, visiting, loung-
ing, sleep.*
*Earthenware jar for water, covered in gunnysack, hung on
Ramada west wall with enamelware mug.*
Wire or sotol baskets: eggs, vegetables, dried ceremonial corn.

The standing of a Kickapoo woman is bound up in her role as
house owner, builder and maintainer. Houses belong to the
family's oldest woman; it is her responsibility to call upon
family and clan folk to help her during the New Year ceremonial enlargement of the *wikiup*, or during the fall and
spring dedication rites when the roofing mats are ceremonially
transferred and new fires lit.

These mats are sewn together from cattails which grow
nearly 200 miles away from their Mexican village. First the
Mexican Kickapoo cattail gatherers say prayers and burn
Indian tobacco before harvesting. At home the bundles of
cattail leaves are thinned so only nine foot lengths remain;
these are then soaked and laid on the ground beside each
other. Teams of women work to sew the 18 by 9 foot mats
with long needles, iron or bone, with *pita* (yucca-like fiber)
or commercial hemp. Bench mats for sleeping and sitting are
hand-woven from dyed and softened Sotol leaves.

This material is excerpted from the forthcoming book
Native American Houses, *by Peter Nabokov and Bob Easton,
published by Oxford University Press.*
Sources for information appearing in this article: The Mexican Kickapoo Indians, *by Felipe A. and Dolores L. Latorre
(University of Texas, Austin, 1975) and* The Mexican Kickapoo, *by R. E. Ritzenthaler and F. A. Peterson (Publications
in Anthropology, Public Museum of the City of Milwaukee,
No. II, 1956).*

The summer house frame is constructed of one-seed juniper
or desert willow. The walls are vertical, the roof frame lattice
is hipped and extends about two feet beyond the *sotol* cane
walls. The first two mats laid at the ends run north-south,
while this second layer runs east-west.

The completed summer house has no smokehole; the
smoke seeps through the sides and eaves. Since these women
are building early in the season, canvas covering is added to
assure warmth.

Cookhouse
12' square
Framing and construction similar
to summer house

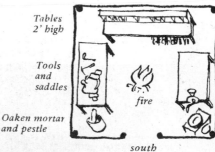

*Cardboard wikiups beneath the international bridge at Eagle
Pass, Texas in 1978, built by a 600-member Kickapoo tribe
that has asked the U.S. federal government for a $1.5 million
grant to establish a reservation where they can practice their
religious rites and educate their own children.* □

UPI Telephoto

North American Houses and Barns

Sod Houses

by Roger Welsch

. . . For thirty years the standard on the plains was the sod house. (How sod came to be used as a building material is uncertain. The settlers may have borrowed the idea from the Mormons, who began building with sod in the mid-1850's. The Mormons, in turn, probably got the idea from the earth lodges of the Omaha and Pawnee Indians.)

Today we make every effort to design houses that bring the out-of-doors indoors and take the indoors out-of-doors. Patios serve as dining rooms, huge windows provide the illusion that we are outside when we are inside. We open the house walls and break down barriers. But for nineteenth-century plains dwellers, perceptions were different. After a day of being squeezed between sky and earth, of being exposed to the withering sun or a razor-sharp wind, there was little desire on the part of the pioneers to bring the environment into their houses.

The house was meant to be a fortress, a bastion for shutting out the outside. The thick walls, the few small windows and close rooms were not seen as disadvantages — as they might be now — but rather as an integral part of the sod house's advantages. Far from being discomfited by the cramped quarters, plains settlers sought the closeness of family members in the evening hours, after a day spent out of sight and hearing of each other or, for that matter, of any other human being. The close contact and association with the family took on a very special, desirable quality . . . .

Many women, singly or in groups, homesteaded during the 1800's. Shown here are the Chrisman sisters, Custer County, Nebraska.

John Curry sod house near West Union, Custer County, Nebraska, 1886.

My house it is built out of national soil,
The walls are erected according to Hoyle;
The roof has no pitch, it is level and plain,
But I never get wet—unless it happens to rain.
                    From a Pioneer Plains Folksong

The simplest form of sod house, built low to the ground, with no windows.

. . . As a familiar plains' line goes, "Living in Nebraska is a lot like being hanged: the initial shock is a bit abrupt but once you hang there for a while you sort of get used to it." In the demise of sod houses, the forbidding mystery of the plains had dissipated to the extent that inhabitants no longer felt the need for the physical and psychological security that these dwellings offered.

Sod houses still dot the plains. Some are still lived in, but most are just derelicts — abandoned, their roofs overgrown, their door and window frames sagging. These ghosts, however, are more than merely abandoned houses. They are reminders of the grip the plains had on their early settlers. Behind their dark sod, these houses offered protection from a lonely and inhospitable land. They also offer another reminder — their abandonment and replacement by wood frame houses are symbolic of a reversal in attitude. Now, it appears that plains dwellers have a grip on the land instead of the other way around. Thus the sod house was as much a product of the impact of the plains on the human mind as it was a product of the geography of the plains.

A farmer friend of mine commented a short time ago, "We seem to forget that we may have made this land what it is, but first it made us what we are . . . ."

Reprinted from *Natural History* magazine, May, 1972. □

Old Texas Buildings

by Burton Wilson

58

Washington

Connecticut

Maine

Country Buildings

Nova Scotia

Kentucky

Connecticut

Nova Scotia

Mississippi

California

Barns

The barns of North America are a simple, practical expression of a way of life, the land, and the people who built them. They are living examples of the type building that occurs whenever efficiency, economy and durability are the influences shaping design.

The barn builders were anonymous farmers and carpenters who heeded local weather conditions, understood siting requirements, built with available materials, and let practical purpose rather than style govern design. Often the owner, designer, builder and user were the same person.

Few barns are being built in America these days. Feed is seldom raised and stored on the farm, but is now shipped in from other areas. Barns are going the way of the family farm, which is unable to compete with the (short-term) cut-rate prices of mechanized chemical agriculture. Many are either falling apart, or being burned or torn down. Yet those that do remain, that are maintained by their owners, are graceful and harmonious examples of functional design, of the days of a land-conserving and energy-efficient American agriculture.

continued

*This Washington barn, built in 1925, is 75 feet long, 55'-6"
wide, and 45 feet high at the gable. Rafters are 50 - 55 feet
long, full length cedar poles from old growth, high-altitude*
*trees, chosen for high strength and light weight. The roof is
of hand-split shakes. The entire barn, with the exception of
wall siding, was built without the use of a sawmill.*

These large barns in the state of Washington show the same "aisle and bay-divided" structural framework of the great European barns (see *Shelter*, pp. 30-32). Roof rafters do not span from wall to ridge; rather, there are two interior rows of posts with beams that provide intermediate rafter support.

Incidentally, this type roof structure is what shelters most of Europe's Gothic cathedrals (over the vaulted stone ceilings).

As the European settlers moved west in North America, they framed their barns this way. On the west coast, with light snow loads, large barns were built with milled lumber and on a weight-per-square-foot basis, are among the world's lightest weight buildings.

This barn was built in 1912 on the farm of Willis Chambers near Port Angeles, Washington. Francis Chambers, who was a boy at the time, recalls working on the barn, that the head carpenter was a man named Hartlett, and that it took about three months to complete the building. The barn is 80'-6" long, 76'-6" wide and 43 feet high from ground to ridge. The framing is of all clear cedar.

The large barn, above, down the road from the Chambers barn, is falling apart. When we were looking at it an old man came out, said "What good is it?" Large barns were formerly needed for hay storage in this rich agricultural valley, but modern feedlot livestock practices have made them obsolete.

Silo in Connecticut

Connecticut

Mortise and tenon barn frame, Nov

Above and right: *hay barn in Washington*

Nova Scotia

Washington

Log barn, Washington

Washington

Bungalows

by Renee Kahn
Photographs from David Gebhard

... The term *bungalow* comes from the Hindustani word "Bangla" (literally — from Bengal) and signifies a low house surrounded by porches. These houses were not typical native dwellings, but were the "rest houses" built by the English government in India for the use of foreign travellers. Rambling one storey structures, they were designed to withstand the heat of the Indian climate, and had wide overhanging eaves, stone floors, and long, breeze-filled corridors. Deep verandahs (another Indian word) provided additional shade. The word *bungalow* was brought back to England by retiring civil servants, and eventually came to describe any modest, low-slung residence of picturesque lines.

In the United States, the term *bungalow* supplanted the word *cottage* and was popular because of its euphonious sound and exotic connotations. During its heyday prior to World War I, thousands of bungalows were built ....

Despite wide variations in style, cost and location, the bungalow had certain, almost universal characteristics. Its lines were low and simple, with wide, projecting roofs. It had no second storey (or at most a modest one), large porches (verandahs), and was made of informal materials. It was primarily for use as a summer, or resort house, except in the warm California climate, where it was easily adapted to all year round use. . . .

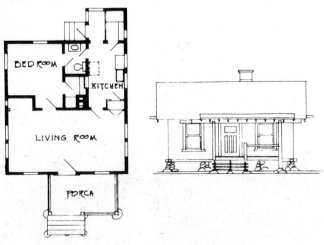

Small speculative house designed by Charles and Henry Greene, California architects, in 1906.

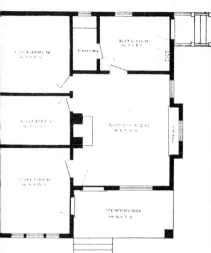

Floor Plan of No. 2009

Bungalow Design No. 2009 Price of Plans and Specifications **$5.00**

Full and complete working plans and specifications will be furnished for $5.00.
Cost of this bungalow is about $900, according to the locality in which it is built.

Porches were an essential part of the bungalow style, but unfortunately, they were designed for sunnier climates, and darkened the interior of the house. This was often overcome by constructing the porch with an open roof, like a trellis, which could be covered by vines or an awning. Porch roofs frequently echoed the gable of the house, but were placed off to one side. Posts were made of boulders, or covered with shingles, contributing to the desired "natural" look. This natural look also extended to the outside wood finish which was either left plain, or stained, sometimes with a lump of asphalt dissolved in hot turpentine. . . .

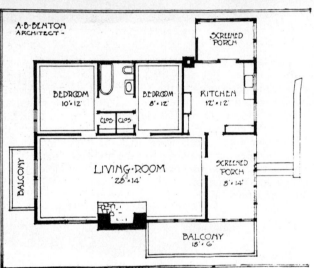

A·B·BENTON
ARCHITECT –

BEDROOM
10'·12'

BEDROOM
8'·12'

KITCHEN
12'·12'

SCREENED
PORCH

CLOS CLOS

BALCONY

LIVING·ROOM
26'·14'

SCREENED
PORCH
8'·14'

BALCONY
18'·6'

The flood of literature after the turn of the century brought much advice on how to furnish the bungalow. Simplicity, and lack of pretension were the main goals. Gustav Stickley, the furniture maker, was also editor of the magazine "The Craftsman," and was one of the major promoters of the bungalow style, which he referred to as "Craftsman Homes." in 1909 he wrote: "When luxury enters in, and a thousand artificial requirements come to be regarded as real needs, the nation is on the brink of degeneration."

Stickley, a disciple of William Morris, was also responsible for the sturdy oak furniture commonly known as "Mission." These comfortable, handcrafted pieces were considered appropriate for the bungalow, as were the plainer versions of wicker and rattan. Easy-to-care for leather or canvas covered the seats. No pretty bric-a-brac lay about, only sturdy art pottery and brass or copper bowls. Matting and shag rugs were suggested for the floor; however, Orientals were "never out of place." Surfaces were simple, and covered with natural looking stains. . . .

Bungalow Design No. 2001

Price of Plans and Specifications $5.00

Floor Plan of No. 2001

Full and complete working plans and specifications will be furnished for $5.00.
Cost of this bungalow is about $700, according to the locality in which it is built.

It seems ironic that the bungalow originally had its greatest impact upon the intellectual upper middle class who valued it for its "honesty" and "practicality." Despite its lofty aspirations and exotic sources, the style ended up sloppily imitated in thousands of tacky boxes. It has come to represent both the best and the worst in American architecture from the turn of the century until the 1920's.

It did, however, make positive contributions to the American home with its lack of pretentiousness, its use of natural materials, and its effort to integrate the house with its surroundings. Its direct descendant, the ranch house, a somewhat characterless version of the bungalow, remains today one of the most popular forms of domestic architecture.

Stickley saw them as " . . . the kind of houses that children will rejoice all their lives to remember as 'home,' and that give a sense of peace and comfort to the tired men who go back to them when the day's work is done." □

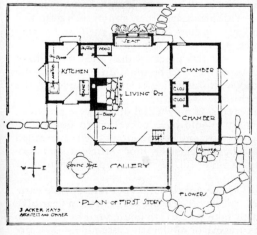

Text reprinted from The Old House Journal, *September, 1977.*

Design

In most situations, the stud frame house is the most practical way to build in North America today. Other forms of building —adobe, stone, logs, post and beam, etc.—may be more appropriate where the materials are available, and time plentiful. But in most cases, conventional stud construction of a small building will be the quickest, cheapest and most durable way to build a home.

In the pages that follow we present a survey of the design and construction of small stud frame buildings. The *Design* section (pp. 74-110) includes information, hints and advice on *design* of a small building, from siting, climatic and planting considerations to a discussion of basic roof shapes, floor plans, and additions and variations to these basic shapes. The *Construction* section (pp. 111-134) is an introduction to *construction* of a small wood-frame building, from foundation to roofing, as well as interior finish and miscellaneous building tips.

This is not a complete design manual, nor a thorough treatment of construction practices. Rather it is an introduction to some of the principles and practices of design and building of small homes.

Advantages of Stud Construction

— most building materials are manufactured for this type construction, and are commonly available.
— rectangular floor plan, vertical walls makes expansion easy.
— conventional roofing materials are cheap and durable.
— stud frame is easy to insulate.
— lends itself well to used materials.
— easy to take apart, reuse materials.
— complies with building codes.
— quickest form of conventional construction.

The Abstract Concept vs. The Hard Reality

Building a house is most likely the biggest thing you will ever attempt — in sheer size, money invested, hours spent, energy exerted. In recent years, a great many first-time builders have made the mistake of building a home from an abstract concept — a dome, a sculptural structure, a primitive or vernacular model — rather than from tried and tested building techniques. Many of these projects took more time and money and were far less practical than they might have been: the reality turned out to be quite different from the initial concept, and with something as large as a building, the mistakes can't be thrown away.

Builders are not fools. If polyhedral shapes were practical, the building industry would have converted to dome construction years ago. If post and beam had not been superceded over 100 years ago by the cheaper, more efficient stud construction, builders would still be putting up frameworks of heavy timber.

In these times, you've got to build your house while earning enough money to live on (few people can take a year off). You can't afford to take forever, to injure your health, or to end up with an expressive or modernistic assemblage that promises nothing but continual discomfort and/or maintenance.

Better one's house be too little one day than too big all the year after.

Thomas Fuller: Gnomologia, *1732*

A very small house is legal under most building codes. You could build a 300 - 400 sq. ft. house, with kitchen and bathroom back-to-back (for easiest plumbing). Then when you have more time and money, you could add on. One story, short spans, simple to build, well insulated, easy to heat.

Even though you build in stages, it's good to plan as much in advance as possible. Should the addition be to the north, the south . . . ? what about roof lines . . . ? how will it tie in . . . ? etc.

To check out a proposed floor plan: draw it out with chalk on a schoolyard or pavement. Include interior partitions, kitchen, furniture. Walk around within it to get a feel of the space. □

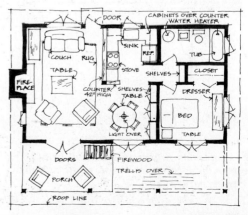

Climate
Site
Planting

A thorough understanding of your site will help you in making sound decisions in placement and orientation of a building. It would be ideal (though not usually possible) to live on the site for a year in a trailer or small shed and watch the changing of seasons, the wind force and direction, the variations in sunshine and rainfall, and temperature fluctuations *before* you build. In lieu of a long period of on-site observation, there are many factors that can be observed and researched that will help in good design. On these two pages are some very brief notes on siting, climatic and planting considerations. This is not a complete list by any means (books are written on each of these subjects), but rather a few basic ideas of these aspects of the design process.

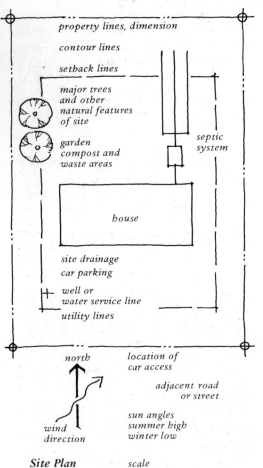

property lines, dimension

contour lines

setback lines

major trees and other natural features of site

garden compost and waste areas

septic system

house

site drainage
car parking

well or
water service line

utility lines

north

wind direction

location of car access

adjacent road or street

sun angles
summer high
winter low

Site Plan   scale

A site plan drawn to scale will help determine location of house, septic system, trees, garden, etc.

Climate
Temperature
— consider daily as well as annual fluctuations.
— determine if there are any likely heating or cooling problems.

Rainfall
— determine rainfall through the seasons.
— what is direction of wind-blown rain?
— drainage around house.

Frost
— footings should be below frost line.
— will pipes freeze in winter?

Snow
— snow loads require heavier roof and foundation.
— snow insulates.
— eave design is important for shedding snow (see p. 123).

Winds
— there are daily and annual variations, often with a regular pattern.
— protection of outside sitting area and garden from strong or dusty winds.
— using pleasant breezes to cool interior spaces.

Sunlight
— indirect sunlight (as from the north) provides an even source of light.
— direct sunlight is useful for warming rooms or solar heating water.
— sun is low in the sky in winter, strikes south walls strongly, penetrates deeply into rooms through south-facing windows.
— sun is high in sky in summer, strikes roofs strongly.

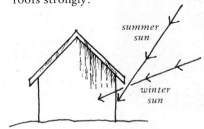

summer sun

winter sun

Site
Microclimate
Whereas *climate* generally refers to the conditions distributed over a large area, *microclimate* refers to the particular climate of one small location: " . . . at ground level multifold minute climates exist side by side, varying sharply with the elevation of a few feet and within the distance of a mile. . . ." * Microclimate can be as important or more important on the site than climate. Microclimatic conditions change faster than climatic. Microclimate is often man-made, especially in cities.

*Design with Climate, Victor Olgyay.

Orientation of slope
— south slopes are warmest; north are darker and cooler.
— eastern slopes get the morning sun; western get afternoon sun, are usually warm from daily heat build-up.

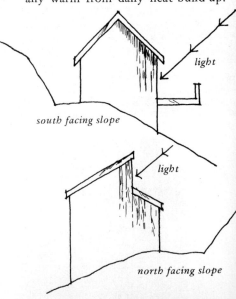

light

south facing slope

light

north facing slope

Existing vegetation
- provides clues to soil, rock and water conditions.
- large trees can disturb foundations, may fall in storms, can provide (welcome) shade from hot summer sun, (unwelcome) shade for vegetable garden.

light

arbor for cool north patio

deciduous shade trees

Soil stability
- condition of soil at surface does not tell you what it is like 4' - 6' under. Dig a hole and see. Look for signs of unstable conditions or slides. If any doubt, consult a soils engineer.

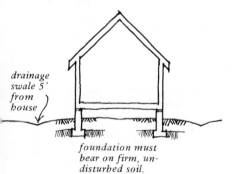

drainage swale 5' from house

foundation must bear on firm, undisturbed soil.

Drainage
- site building for dry footings.
- septic tank should drain away from house, never be above water supply.

View
- windows can afford panoramic or framed view.
- you can make a view with patios or planting.

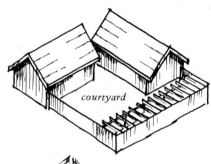

courtyard

Privacy, noise, access
- plan for maximum privacy by orientation of building, placement of doors and windows, etc. For example, it is better to face a picture window towards the backyard than towards a busy street.
- consider noise from neighbors and traffic.
- allow for pedestrian and auto access.

Legal requirements are important: local ordinances, setbacks, front yards, easements, parking requirements, etc.

Utilities: check if they are available and if so, where they enter property. Cost of utilities is often high at a remote site.

Local traditions: ask old-timers.

Local architecture: Look at the small buildings near your site, those that were built 30 - 50 - 75 years ago: barns, sheds, coops, older homes. They were generally built with the economy of necessity, of local materials, suited to local weather. Notice roof pitch and overhang, building shape, orientation, siting, materials, details.

Planting
Plants can provide shade, privacy, wind protection, fencing, fruit, foliage, bird and wildlife protection, etc.
- deciduous trees will provide shade in summer, allow sunlight in winter.
- dwarf fruit trees can produce fruit in three years, can be planted 6' - 8' apart but need more care and feeding than full-size trees.
- evergreens (pines, firs, etc.) can be planted on north side for winter wind break.

- if a vegetable garden is planned, consider its placement in relation to kitchen.
- plants absorb sound.
- a good book on planting: *Plants/People/And Environmental Quality* (see bibliography).

Design Checklist

Checklist of Needs

These are ideas from people who have built their own homes: a random assortment of pre-design thoughts based upon occupants' needs and desires. As with the preceding two pages, this is not a complete list; yet these examples may give you a start in preparing a list for your own individual situation. To begin, we suggest you imagine yourself in a house: going through daily activities, looking out of windows, working in the kitchen, walking from one room to another, etc., as an aid in making up a checklist of your own needs and desires.

Early houses generally had one room for all activities; the fireplace was used for cooking, heating, and evening light. Later, in the industrial age with more abundant materials and fuel available, houses were partitioned into separate rooms for kitchen, dining, living, and sleeping. Now, in an era of dwindling fuel supplies, the earlier design of a one-room kitchen-dining-living space may be an efficient model to consider. A wood-burning stove can cook food while heating the room and its occupants. Cooking on a gas or electric stove will send some heat into the room, and the oven door can be left open after use to catch the remaining heat in the room rather than lose it through the flue. Warm blankets instead of heaters in sleeping rooms. Other rooms for daytime work need less heat if you are moving around, and warmly dressed.

Arrangement of rooms

- small houes are often one big room. Think in terms of areas for related activities rather than individual rooms and of activity centers rather than shuffling tiny rooms around. A small house isn't simply a scaled-down big house.
- see pp. 88-101.

Main room

- you can start small, building one main room with a kitchen. Later, as time allows, and finances permit, other rooms can be added. The main room can be basic living space, a base of operations for finishing the job. Thus it is important to plan as much as possible before you begin, considering the site and future additions in relation to this first phase.

Circulation

Crossing a room at a diagonal wastes space; try to have circulation along the perimeter.

Either minimize circulation space, or exploit it by enlarging somewhat to accommodate another activity.

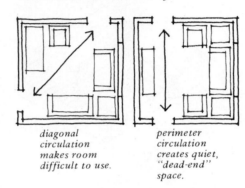

diagonal circulation makes room difficult to use.

perimeter circulation creates quiet, "dead-end" space.

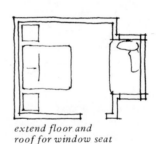

extend floor and roof for window seat

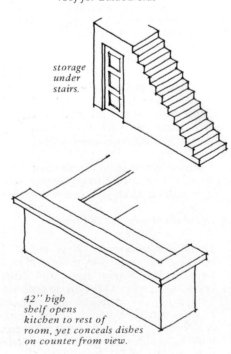

storage under stairs.

42" high shelf opens kitchen to rest of room, yet conceals dishes on counter from view.

Kitchen

- enough counter and storage area.
- sloped drainage counter so dishes can drain into sink.
- good working relationship between sink, stove, refrigerator.
- toe space under counters is important.
- counter height: standard is 36", but many kitchens have lower counters for kneading bread or chopping food.
- morning sun on kitchen table (S.E. window for low winter morning sun).
- counter height electric receptacles.

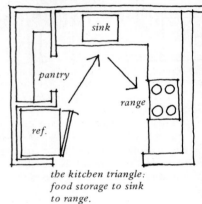

the kitchen triangle: food storage to sink to range.

Utility room

- a small room off the kitchen can be used for removing wet or muddy clothes or boots, or can be utilized for food storage, or for a washing machine.

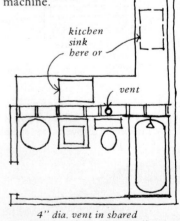

4" dia. vent in shared 2x6 stud plumbing wall.

Bathroom

- fixtures aligned in row on 2 x 6 stud wall is most efficient.
- hot water heater in bathroom helps heat room, is good place to dry towels over.
- best plumbing arrangement is with bathroom, kitchen back-to-back.

Sleeping

- bedrooms use a lot of space, which may be more useful elsewhere in the house.
- children spend more time in their rooms than adults; maybe they should have larger rooms than parents.
- sleeping areas in lofts, alcoves off the living room, or merely a corner of the main room.

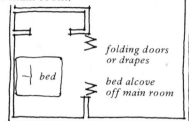

folding doors or drapes

bed

bed alcove off main room

Storage

- some storage should be more accessible than others.
- closed storage areas for less frequently used items, open storage for items in daily use.
- roofed-over outside space for firewood, bicycles, etc.
- porch or mud room for shedding muddy boots, rain gear before entering house.

Basements

- cool space for storage.
- should only be built on sites with low water table and/or good drainage.

Workshop

- for home maintenance, tools, etc.

Doors

- doors generally open inwards; if door swings out, hinges are on outside and pins can be removed to break into house.
- porches or overhangs protect exterior doors from rain and wind.
- try not to place doors on windward side of house.

Let all the principal chambers of *Delight*, All *Studies* and *Libraries*, be towards the *East*: For the Morning is a friend to the Muses. All Offices that require heat, as *Kitchins, Stillatories, Stoves*, roomes for *Baking, Brewing, Washing*, or the like, would be *Meridionall*. All that need a coole and fresh temper, as *Cellers, Pantries, Butteries, Granaries*, to the *North*. . . .

The Elements of Architecture,
Sir Henry Wotton, 1624

Windows

- view.
- ventilation: screened openings keep out flies.
- light.
- sitting areas that receive sun at different times of the year.
- windows so you can see someone approaching house before they arrive, for keeping an eye on children at play.

vine on trellis for sunshade

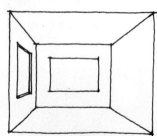

windows on two walls balance light, help ventilation.

Porches

- south facing porch for winter warmth, north facing for summer shade.
- large porch by main entrance provides storage for firewood, place to hang clothes, stay dry while opening door.

Noise

- bathroom between bedroom and living room.
- closets, bookshelves between rooms.

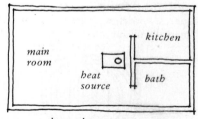

closets between bedrooms

Heat source

- heat source in relation to living pattern: for example, plan nighttime living area close to heat.
- small rooms heat easily.
- large rooms generate drafts.

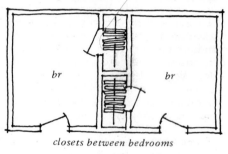

kitchen

main room

heat source

bath

locate heat source near center of building

Available materials

Many of the decisions about house design and construction will be based upon the materials available locally. Used lumber, doors, windows are usually for sale in or near cities. In the northeast U.S. new Spruce is available from nearby sources. In the Pacific Northwest Douglas Fir is a local material. Many areas have stone; adobe or rammed earth is suitable in dry regions. A careful analysis of locally available materials is an important factor in house design. A study of materials used in nearby buildings can provide helpful guidelines. □

Flows

In the past few pages we have considered the physical characteristics of a house — its shape, placement on the site, relation to the elements — as well as functions and use of interior space. Now we will look at another group of characteristics about a home: the *flows* that move through a house.

Clean water flows in, waste water flows out. Cold and hot water are moved around in the house. Fuel comes in; heat circulates, as do breezes. Food is brought in, compost, garbage and human wastes move out. Dirt and dust come in and are taken out. Smells move about. Electricity flows in and is converted to heat, light and power . . .

These movements of liquids, materials, warm and cool air, and energy are all important to consider in the planning and design of a home. What to do with wastes, where to place a heat source, where the food comes in, how and where food scraps are disposed of, where electricity or gas and water come in, how natural breezes can cool the house . . .

Following are a few brief ideas on some of the flows through a house:

Ventilation-cooling
— *air flow:* windows, dormers, belvederes, porches, balconies can help carry heat out.

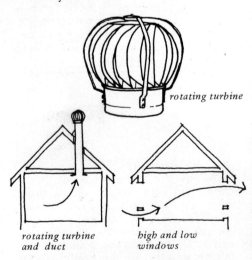

rotating turbine

rotating turbine and duct   *high and low windows*

— *buffer rooms:* porches, mudrooms, verandas modify heat gains and losses.
— *shading:* overhang, trellises can keep summer sun out.
— trees can deflect air currents around a house, or can be sited to increase wind flow towards a house.

Heating
— plan winter living near heat source.
— drapes, shutters, storm windows can help conserve heat. *Portieres* are heavy drapes hung in doorways to retain heat.

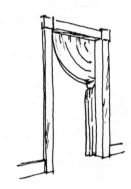

— orientation of windows (southern exposure) allows heat energy to flow in; sunshine is a "fuel."
— weatherstripping around doors and windows keeps cold out, warmth in.
— insulation; see p. 129.

Water Heating
— hot water accounts for a great deal of the average household's energy use.
— locate heater close to fixtures (kitchen and bath).
— pressure relief valve on hot water heater is important, can prevent explosion.
— conserving hot water: a) turn thermostat down b) insulate tank and pipes with fiberglass.

Waste Water Out
— waste water contains soap, grease, food particles etc. that can damage plants, compact soil, destroy microlife when applied directly to soil.
— reclaimed "grey water" (drain water only) can be used for irrigation if properly filtered, or digested in small septic system to produce clear effluent.
— boron is destructive to plants.
— bathtub water is fairly clean for use on plants.

Odors
— windows, hood, or pop-up vent over stove will exhaust cooking odors directly.
— openable window next to toilet.

Garbage
— area to separate glass, paper, metals for recycling.

Food In and Out
— consider where vegetables from garden, groceries will be brought in, how they will be stored (crocks, tins, jars, etc.).
— food scraps can be kept in plastic bucket, emptied every few days, or smaller container and emptied each day.
— compost can be worked into compost pile, or buried in trench.

People In and Out
— an entry porch can be a place to take off wet rain gear, muddy shoes.
— door mats — rubber on the ground level, fiber on the doorstep — can help keep out dirt.

Drying Clothes
— outdoor clothes line close to washing machine.
— indoor clothes drying: hang nylon strings 8" apart behind fireplace, stove, furnace or in loft for drying clothes on damp days. □

Toilets

Human Waste		
Flush Toilets	*Water*	*Approx.*
Type	*Usage*	*cost 1979*
standard flush	5 - 7 gal.	$80-100
water saver	3½ gal.	$75-100
water/comp. air*	2 qts.	$265†

*For gas stations, public restrooms, etc. Shown here as indication of design advances in flush toilets.
†Plus air compresser.

Conventional toilets use a great quantity of clean drinking water for flushing; unless a septic tank is used, water-borne sewage is expensive to treat and often environmentally destructive. However, there are definite advantages: they are easy to use, can be located anywhere in the house, feces are not exposed to insects, and even pathogen-carrying feces are safely disposed of.

Biological Toilets, Composting Privies

Where municipal sewage systems are not available, development has been limited to sites where soil percolation allows for septic tank disposal of domestic wastes. Small communities as well as large cities are now required by federal regulations to "sewer up," or upgrade existing facilities. These factors, plus water shortages, have led to heavy criticism in recent years of water-borne systems.

Several types of waterless toilets have recently come to public attention. One type is the biological toilet; best known is the Clivus-Multrum, a large fiberglass chamber used as toilet and for food scraps disposal. There are ventilating pipes inside, and gradual decomposition in the downward sloping chamber is supposed to eventually produce clean compost. Another type is the composting privy, such as the Farallones Composting Privy: a concrete block structure with two four-foot square chambers. Sawdust is added after each use; when one chamber is full, the other is used. Piles are usually turned with a pitchfork.

The theory of a composting toilet is that human wastes, along with a carbonaceous material such as wood shavings will eventually be converted into a rich compost of 10-20% its original volume when placed in a well-ventilated chamber. These toilets rely on aerobic (with oxygen) decomposition, unlike outhouses. Like compost piles, the process must be a living system, with balanced food, air, moisture and warmth.

Problems with composting toilets and privies:

Epidemics such as hepatitis, cholera, typhoid fever, and dysentery have been caused from viruses of human waste origin. The spread of parasites such as hookworm, whip worm and Ascaris is also possible. Thus, health agencies have been reluctant to approve these units for wide usuage. Both the Clivus and the Farallones Privy have had problems. Fruit flies (via food scraps) and manure flies have been known to infest Clivus toilets; once established they are difficult to get rid of. Several tropical parasites (including hookworm) were recently found in a California composting privy after a six months composting period.

Specific problems:

— temperatures must reach 160° for a sustained period to kill parasitic ova; this is often not the case.

— lack of carbonaceous material makes the pile solid, does not allow the necessary ventilation. Or, clogged air intakes or ducts or insufficient updraft prevents aeration.

— fly infestation, which can cause spread of pathogens.

In a recent research report on compost toilets prepared for Rodale Press, Patti Nesbitt concludes:

Disposal of wastes in a contained tank, even if the wastes do not undergo the decomposition necessary to inactivate all enteric organisms, is still a step forward in terms of water conservation and the reduction of water pollution.

These systems, though, should not be advanced for their potential to create soil conditioner. It is likely that, for a few years to come, most composting toilets will produce contaminated humus. Even if the center of the pile reaches 160°, the fringes may not. It is recommended that the compost . . . be buried twelve inches down for two years to insure its purification . . . .

It is possible that compost toilets may never be applicable for every home; they are most suitable (especially the large tank units) in low-density areas with sufficient space for the tank and burying the humus . . . . We are concerned about the lack of rigor and the jump-on-the-bandwagon attitude being taken by some in drought-stricken California. The creation of potential health hazards is not a step in the direction of solving the problem of sewage treatment and disposal. It is time to admit openly the present limits of these alternative systems and to begin serious work to make them viable and more generally applicable.

Water and Waste Information:
Building Your Own Compost Toilet and Greywater System, by Zandy Clark and Steve Tibbits. $3.00 from Alternative Waste Treatment Association, Star Rt. 3, Bath, Maine, 04530. Best publication on compost toilets.
Residential Water Conservation, Report No. 35, 1976, California Water Resources Center, UC Davis, Davis, Ca. Good review of conservation devices and where to get them.
Small Scale Waste Management Project, by Robert Seigriest, Dept. of Environmental and Civil Engineering, U. of Wisconsin, Madison, Wis. Send $1 for list of publications. Excellent design papers on mounds, slow sand filters and greywater. □

Alternative Energy

We want to believe in breakthrough solutions. We would rather hear about solar heaters or wind generators than about reducing wasteful consumption; about revolutionary housing design instead of making better use of conventional construction methods. We would like to believe there are *alternative* sources of energy to maintain the present American standard of living.

In the last decade, especially since the oil embargo, we have seen increasing interest in what are popularly called alternative energy devices: solar heaters, wind generators, methane digesters, etc. Almost all the media coverage — newspapers, magazines, books, television — has been optimistically favorable. Do all these devices work as well as we are being told? Can we rely on the accuracy and objectivity of all the books now available on solar heating, or claims made by inventors in articles on their own discoveries? Our guess is no, and that although many of the devices are useful, even inspiring, a lack of critical analysis by reporters and writers, as well as by inventors and promoters, is not giving the public an accurate picture. A few examples:

Solar space heating (as opposed to water heating): many homeowners are being sold high-priced, complex pieces of hardware to replace oil furnaces; or a conventional heating system is installed along with an active solar system for long cloudy spells — saving on fuel bills, but requiring a high investment for two heating systems. In many cases, building a small house, remodeling, landscaping, wearing warmer clothes indoors, insulation, window alterations, storm sash, could save more energy than buying new, expensive hardware.

Wind generators are expensive, high maintenance machines that produce very small amounts of power in proportion to their cost.

Methane digesters require large amounts of manure; sludge handling is a major problem; they often show a net energy loss, and have been known to explode.

We are not qualified to analyze any of these devices. Nor do we wish to imply that none of them are useful. Yet we believe there is a need for more critical evaluation by the press, more objective analysis and disclosure by inventors and promoters, and a more watchful eye on

the part of the public. It could well do more harm than good for people to believe and invest in devices purported to save fossil fuels or conserve electrical energy, only to find later that the devices fail to perform as expected, or that there are hidden costs or high maintenance requirements.

What is a Kilowatt hour?

Every electric appliance is rated in watts: this is usually found on the name plate of the appliance.

> Model No.
> 120 v–10 watts

1000 watts = one kilowatt (KW). To find out how many kilowatt hours (KWH) each appliance uses, the formula is:

$$\frac{watts \times hours\ used}{1000} = KWH$$

For example: 100 watt light bulb in use for 10 hours is:

$$\frac{100 \times 10}{1000} = 1\ KWH$$

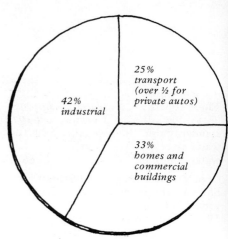

U.S. energy use

- 42% industrial
- 25% transport (over ½ for private autos)
- 33% homes and commercial buildings

Own Your Own Power Company

Vigilante Electric Cooperative, Inc. in Dillon, Montana is an electric company owned by its customers. A non-profit cooperative corporation, Vigilante is typical of many rural electric co-ops in the U.S., with no profits going to pay stockholder dividends or for utility company advertising. For more information: National Rural Electric Cooperative Association, 2000 Florida Ave. N.W., Washington, D.C. □

Working with the Sun
Sun's Angle
Altitude of Noon Sun in degrees (⁰)

Latitude	winter solstice Dec. 21	spring equinox March 21	summer solstice June 21	autumn equinox Sept. 21
25⁰ N (Miami)	41	65	88	65
33⁰ N (Phoenix)	35	57	81	57
42⁰ N (Boston)	25	48	71	48
47⁰ N (Seattle)	19	43	66	43

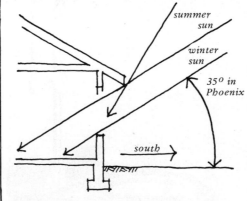

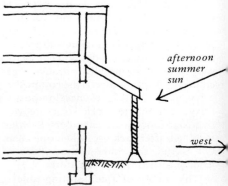

South-facing windows
Eaves or awnings can be designed to screen summer sun, admit winter sun. In warm southern climates, more overhang is required to prevent mid-day heat build-up in spring and fall.

East and west-facing windows
Hot summer afternoon sun strikes west-facing windows. Protection can be provided by trellises, porches, vertical screens, trees, or hedges. East windows can also be a problem in hot summer areas with early morning heat build-up. Hedges away from the house allow more breezes to enter than porches.

Local Energy

The key to having a sufficient supply of energy is to detach one productive process after another from the corporate power network, and restore them to the identifiable human communities capable of actively utilizing sun power and plant power, man power and mind power . . . . Lewis Mumford

Local energy as opposed to alternative energy or world game strategy. Hometown energy. Community or neighborhood energy. Production of basic necessities — food, shelter, heat/power/mechanics — as close to home as possible as a means of increasing our efficiency, our freedom, our independence from centralized control.

Self-sufficiency is a direction, not a goal. Obviously no one can be completely self-sufficient. We all rely on the complex interrelationship of people, organization, production, goods and services to live our lives. But this is no reason not to continue working toward *increased* self-sufficiency: providing an ever-greater share of our own needs. In this quest we begin first with ourselves, then our families and/or friends, then neighbors, then others in our community . . .

Improve Upon What Exists:
— insulate.
— add southern glass, eliminate large areas of northern glass.
— heavy drapes in doorways and windows.
— weatherstripping of windows and doors.
— double glazing or storm windows.
— use wind and sun for drying clothes instead of electric dryer.
— dress warmly inside, keep winter temperature as low as possible.
— use natural ventilation instead of air conditioning.
— water-saver shower head cuts water flow from six to three gallons per minute.
— insulate hot water heaters. Feel the side of the tank; if it is warm, insulation will help. Special blanket insulation is made for this purpose.
— hot water pipes can be wrapped with flexible insulation (available at hardware stores).
— turn off pilot light on gas stove; use welder's sparker to start burners.

What Can Be Done to a Suburban Tract House:
Many of these houses were hastily built for medium cost housing, but the basic cores are often sound — foundation, electrical and plumbing hookups, etc.
— add porches, buffer rooms, trellises for wind control, solar shading, privacy and aesthetics.
— add sunroom or greenhouse.
— remove grass, plant vegetables; plant fruit trees; plant grapes on arbors.
— suburbs are good places to organize cooperative schools and food buying, work projects, community gardens.

Louvered roof and latticed arches shade west-facing deck from direct sunlight, improve house appearance. .

Some definitions
en.er.gy/*n* . . . 1: vitality of expression. 2: the capacity of acting. 3: power forcefully exerted. 4: the capacity of doing work.
self-suf.fi.cient/*adj* . . . 1: able to maintain oneself without outside aid: capable of providing for one's own needs . . . .
econ.o.my/*n* . . . thrifty use of material resources . . . the efficient and sparing use of the means available for the end proposed . . .
in.dig.e.nous/*adj* . . . produced, growing, or living naturally in a particular region or environment . . . .

Hidden Costs of Country Living:
A remote site in the country often involves considerable hidden costs: a well; graveled roads; electric power; difficulties in getting materials to the site, etc. Further, you usually must be a jack-of-all-trades and maintain a lot of machinery in order to survive. Living in a community, on the other hand, whether it be city neighborhood, suburb, or small town, saves energy by being close to services, jobs, markets, schools, and other people.

Indigenous Efficiency
We can learn from the Indians, or from other indigenous people who did (do) not have the vast resources of the industrialized nations (see pp. 4-53). Not that we must migrate with camels, live in wikiups, or till the fields all day. But we can learn much that will be useful in our present lives by observing how efficient people can be when they stay in one place (or migrate along the same routes), when they must work with what is available, when they must heed weather, growing seasons and local traditions. □

Cold Climate

by Ned Cherry

There are a great many factors to be considered when thinking about the provision of shelter in a cold climate. Having spent most of my adult life in the relative comfort of urban apartment houses with each unit usually surrounded top, bottom and each side by other heated apartments, I was rarely inclined to worry about heat. But two years ago, my family and I moved to an old farmhouse in upstate New York where we have just completed two of the most severe winters in recent memory. Trial by ordeal has a way of forcing one to think in the most basic of terms and living in a previously uninsulated 1840 wooden frame house, relying primarily on wood heat, being exposed on all sides to the elements, has been the cause for considerable thought along with the question: why? Having been formally educated in architecture in an urban setting at a time when too much thought about isolated single family private dwellings was considered socially irresponsible, one didn't really give much thought to the problems faced by people living this way. But the fact is, most Americans, rich and poor alike, live in such dwellings, and it is apparent that in most cases little consideration has been given to responding to year-round weather conditions. Following are some thoughts on design and building a home in an area with sub-zero winters; this is not intended to be the definitive word on the subject but rather some ideas based on experience, fact, hearsay and common sense to stimulate further thinking.

Orientation

If you take into account sun orientation and prevailing wind direction from the outset of a building project, half of the problem has been dealt with. Knowing the orientation of winter sun and utilizing it to maximum advantage, and minimizing exposure to cold winter winds by proper siting of the building should be obvious. Yet it is amazing how many houses are affected by view, relationship to road, insensitive building and zoning codes and the like. Proper orientation with regard to the severest anticipated winter conditions is of primary importance and is a decision that should be made before all others. Of course, in acquiring an existing house, one can usually make alterations or improvements. Reducing or eliminating northern facing windows, adding more windows to a south wall, planting trees for windscreens are some possibilities. Even building a wall, an outbuilding, or garage can provide a shield from prevailing winter winds.

Insulation

Insulation is currently a subject of controversy among the building trades with someone finding fault in nearly all the available materials. Fiberglass insulation, although relatively effective in performance, can be irritating to work with and may cause harmful side effects if not handled properly. It has an "R" value of a little over 3 per inch; 6" in walls and 10" in ceilings should prove adequate in most dwellings. Blown cellulose (shredded newsprint blown under pressure) is attractive ecologically, but recently there have been some problems with its flammability. It is fireproofed with boric acid which is now in short supply, may have a tendency to disintegrate in time (depending on application) and will eventually settle, thus reducing its efficiency. Its "R" value is something over 4 per inch. A relatively new foam product called ureaformaldehyde is claimed to have an "R" value of 5 per inch, but is said to shrink in time, thus causing leaks; it is also known to emit noxious fumes in a downward direction after a while, thus rendering its use in ceilings or attic spaces questionable. There are also many types of rigid foam insulations available, the most popular being urethane with an R5 per inch rating. It is, however, derived from the petrochemical industry and produces poisonous gases when exposed to fire. Bearing in mind the shortcomings in each, I tend to prefer fiberglass insulation provided that a good face mask and gloves are worn during installation.

Many builders in the east are beginning to frame wood houses with 2"x 6" studs at 24" on center instead of 2x4's at 16" on center in order to provide the extra space for wall insulation. This is probably a good idea but makes one wonder why frame all walls to the same thickness when it might make more sense to use 2x8's or even 2x10's on northern facing walls and 2x4's on south facing walls where insulation is not as critical. This idea goes back to the question raised a few years ago about the rationale of glass skyscrapers with all four walls of the same construction. At any rate, I believe that the whole field of building insulation is in about the same area of conjecture as that of selecting the right type of solar heating system or collector panel. It may be too soon to effectively evaluate the adequacy of any method. In the meantime, piling up snow or hay bales around the base of a house is an age old remedy for reducing cold and draft penetration.

Windows

Having experienced these past two winters, I would be inclined to discourage anyone from installing windows in a north facing wall and encourage the use of as many southerly facing windows as economically and aesthetically feasible. There's nothing nicer than a bit of winter sun coming into a room, and a south facing window will allow this to happen about 6-8 hours per day even in December. "Thermal pane" windows or windows with two pieces of glass separated by a vacuum or an airspace for insulating value are very effective but expensive, if factory made. Storm windows are a must and I have found that even adding an additional sheet of glass over a fixed window when an odd sized storm window was not available was a big help in keeping out the cold drafts. Obviously, heavy curtains or drapes, along with

Thermal Mass

by Michael Gaspers

shutters, can be effective in keeping in the heat at night as well as cutting down on drafts. And there is nothing like well caulked windows and storm windows to help reduce drafts and heat leakage. With respect to fabricating your own double or even triple glazing, I have not seen too much problem with condensation, at least not so much that it won't usually dry out in the next day's sun.

Roofs

The proper pitch for a roof in a cold snowy climate is 5 in 12, that is 5" vertical for every 12" horizontal. This pitch will allow proper drainage of rain and snow in most cases. Some people prefer the placement of a 24" strip of metal at the edge of a pitched roof so snow will slide off and ice will not accumulate at the edge, causing an "ice dam" and forcing moisture back up underneath the shingles. Others prefer a roof with less pitch, thus allowing snow to remain on the roof throughout the winter and acting as an insulator. This is fine until a midwinter or spring thaw comes, increasing the possibilities of a leaking roof. The old New England "salt box" design with a long sloping roof and short wall facing north to minimize the effects of severe north winds, and a high south facing wall with minimal roof for maximum sun exposure was a sensible cold climate design.

The short north wall could be used for closets to insulate from the north cold. The north-facing sloped roof usually retained the snow for insulation, and the higher south-facing wall could have maximum fenestration to allow the winter sun to naturally warm the house.

Foundations

The important thing in foundation and footing design in cold climates is to be sure your footing is below the frost line, that is, the depth at which the ground ceases to freeze. This point varies from area to area but in most northeastern states it is around four feet. Setting the footing just below this level will insure that your foundation will not move when the ground above freezes, thaws and consequently heaves, thus knocking your foundation all out of alignment. There are various ways to insulate and protect a foundation wall. Most foundation walls these days are built with concrete or cinder blocks and can either be lined on the inside with rigid cellular glass insulation or the cores of the blocks can be filled with either a pre-

molded filler or loose insulating material like "zonalite." Any crawl spaces in the house should be insulated between the joists or beams and some sort of vapor barrier should be put over the ground in the crawl space to help keep moisture out of the wood structure above, as well as reducing cold.

Heating

The correct choice of heating systems these days is really up in the air. Anyone currently building a house should be skeptical about getting tied into or over reliant on oil, natural gas or LP gas. On the other hand, I wouldn't recommend too much reliance on solar heating yet either, except perhaps for domestic hot water use. There are various ideas being developed using passive solar heating techniques such as heat walls, heat windows, sophisticated petrochemical type products, etc. Combination wood-oil furnaces seem fairly good, though expensive, thus allowing some flexibility. I would recommend circulating hot water over forced hot air in any case for a more comfortable, evenly distributed heat supply. Combination wood-coal burners might be the best bet at this point. Heating solely with wood is a nice idea if you've got the wood lot and a dependable chain saw. That's quite an investment. There are still areas in the east where you can buy a face cord of wood (4'x 8' x 12, 14 or 16") for under $20, in which case it still pays to buy it. There are many good wood stoves available now with excellent combustion design. Just make sure it's made with either cast iron or heavy gauge steel. If you're thinking of building a fireplace, make sure you design it to *bring in outside air* so you're not using heating room air for combustion. Heatilators, circulating water loops through your fire box, etc. are all good concepts if installed and designed correctly. If a house is really designed properly in terms of orientation, is insulated adequately and built well with minimal air leaks, is sensibly fenestrated with more regard to the sun and less to views and aesthetics, heating this dwelling should almost be of secondary importance. This may sound somewhat idealistic, but it's really true. Too little thought has gone into these problems in the past and heating costs have rarely been the major consideration in this country that they are now. All of this seems in retrospect to be just common sense and this is what it takes. □

Thermal mass refers to the potential heat storage capacity of bulk materials. Dense or heavy materials usually store more heat than light ones. The following table shows some common building materials, densities, and heat storage capacities.

Material	Density lbs./cu. ft.	Specific Heat*	Storage Capacity**
Water	62	1.00	62.5
Iron	490	0.12	59
Concrete	140	0.23	32
Stone	170	0.21	36
Adobe	100	0.22	22
Sheetrock	50	0.27	13
Wood	30-40	0.30	9-12

* Btu/lb./°F
**Btu/cu. ft./°F

Note: Btu = heat required to raise one pound of water one degree Fahrenheit.

Water stores more heat by weight or volume than any common material. Concrete and stone are next best, followed by adobe, sheetrock and wood, which stores the least heat.

All of these materials are poor insulators (resistance to heat flow) except wood, which is a moderately good insulator.

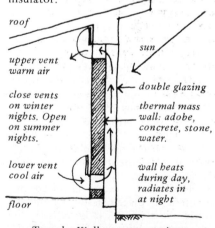

Trombe Wall — cross section from Homegrown Sundwellings

A thermal mass has the effect of absorbing heat as the house warms up, releasing heat as the house cools, thereby smoothing out or regulating room temperatures. Also, these masses soak up and store excess heat, which would normally be vented out of a house and wasted.

Thermal mass can be effectively utilized in home design by building a well-insulated shell with thermal mass (concrete slabs, stone fireplaces, stone floors, water containers, etc.) *inside* the insulated shell. A house thus built requires less heat input, and will maintain more

continued

Building Costs

Thermal Mass *continued*
constant temperatures through the day/ night cycle. Some good traditional examples of houses with large thermal mass are the adobe buildings of the Southwest, and the sod houses of the Prairie.

One simple way to utilize thermal mass is to construct an insulated masonry floor (concrete, stone, brick). For best heat absorption, the floor should be a dark color and exposed to direct winter sun through an adequate amount of south-facing glass. Night heat loss through the glass is prevented by double glazing, tight shutters, or an insulating cover.

Another approach is to make the south wall a large thermal mass, such as concrete or stone, covered with glass (the Trombe wall), or stored water (55 gallon drums or water-filled culvert pipes). Again, these walls must be double glazed or provided with moveable covers.

Thermal mass has also been placed in roofs with insulating covers which allows for both heating and cooling of the rooms beneath, depending on whether the covers are removed during the day or at night.

These designs utilize integral parts of the house to achieve solar heating without the use of expensive, complex, mechanical and electrical equipment.

Caution: Building walls of concrete, stone, adobe, or any massive material can be dangerous in earthquake zones. Be sure such structures are designed accordingly.

So-called "active" solar heating systems also use thermal mass as the heat storage medium. The heat storage material (rocks and water tanks are most common) is placed in the basement or under the floor. Heat is transferred to the storage material from a separate solar collector by circulating air or water. If the collector is above the storage, as on the roof, pumps or fans with associated controls are required. If below the storage, as on a south-facing slope, heated air or water in the collector will thermosyphon up into the storage medium. Active systems are usually more complex and expensive than passive types, but have the advantage of storing heat longer and being able to release heat only when needed.

A good source of information on thermal mass and sun-tempered houses is *Homegrown Sundwellings*, by Peter Van Dresser *(see bibliography)*. □

Borrowing vs. Pay-As-You-Build

Borrowing money from a lending institution to either purchase a home or finance construction involves a total expenditure many people are not aware of. For example, a $50,000 home, with a $10,000 down payment and a 25 year loan on the balance at 9% interest, means monthly payments of $335.00. Over the 25 years, the borrower will pay back $100,500 on the $40,000 loan. (By the time you add taxes, maintenance and insurance, the monthly figure will be closer to $600, or $180,000 over the 25 year period.)

Since the cost of a building is generally equally divided between labor and materials, the costs for this $50,000 house are, roughly:
- materials: $25,000
- labor: $25,000
- bank interest: $60,000

One might well ask what the lending institution has done to deserve more money than the cost of both labor and materials. Surely the processing of a loan does not involve nearly as much labor as the carpenters, plumbers, electricians and other workers contribute. Nor does the handling of a loan seem to consist of such real value as the manufacture and delivery of the tons of concrete, wood, roofing, appliances, plumbing and wiring fixtures that make up the house. One might ask further if such a disproportionate share claimed by banks might not be one of the primary causes of the escalating housing problems in America today.

One obvious way to escape such a long-term financial obligation is to pay for a home as it is built. If loans are necessary, they will cost less if they are in small amounts, for short time periods.

To pay for a house as it is built, it is necessary to carefully estimate costs of each phase of construction, in advance. Since money will most likely be available in increments, it is well to plan where to stop building when the money will run out. Two phases of construction seem ideal: completion of the foundation work and concrete pour, and the framing and roofing of the structure. At each of these points, work could stop until more money becomes available. An example of a bad place to halt construction would be with the floor on, no roof, and winter rains coming.

Paying for a home as you go will involve sacrifices in the early years; often working at a full-time job with all spare time devoted to building. Later, it may mean camping in an unfinished shell, sawdust and sheet rock tracks in your living space, and months, even years of finishing up bit by bit. Yet in the long run, the freedom from an enormous obligation to a bank may be well worth the hardship and inconvenience.

Acting as Your Own Contractor

A contractor performs many valuable functions during construction of a house: attending to necessary permits, negotiating with subcontractors, and assuming responsibility for performance and completion of the job. Many people may find it worthwhile to engage an honest contractor to either assist or supervise their home-building project.

But if you do not have the money, and/or want to be as closely involved with all phases of the job and perform as much of the work yourself as possible, you may want to act as your own contractor.

A contractor solicits bids from subcontractors for various phases of the job. You may want to do this for certain difficult jobs, such as foundation work, plumbing, or electrical installation. Always get at least three bids. Try to check on the quality of work and dependability of the subcontractors.

Prevailing Union Wages in U.S. in 1978
(Cost to Employer)

Carpenter	$16.24 *per hour*
Electrician	16.32
Laborer	12.29
Painter	14.79
Plumber	17.23
Tile Setter	15.77

Used Materials

Used materials and free materials such as stone or adobe save money, but take time. Used wood requires more time than new lumber; if it does not have to be de-nailed, it generally must be cleaned (especially if used inside), and it will take more time to utilize the varying sizes than new lumber, which can be purchased any length.

The amount of free or used material to be utilized is generally a matter of finding the right balance among time available, money to be spent, and how quickly your home must be built. □

Small
Buildings

Each part of the country has special characteristics that favor particular roof shapes. Snow, rain, heat, cold, available materials, practical usefulness, and local experience all contribute to determining the final shape of a building. There are flat roofs in New Mexico, steep gables in Washington, "saltbox" shapes in New England, gambrel roofs in Nova Scotia and hip roofs in Mississippi.

A *flat* roof is usually found where there is little snow and low rainfall, often on adobe buildings. A *shed* shape is a flat roof tilted up, still one plane; it sheds water better than a flat roof. The *low gable* is the most common shape in America today, the most practical shape for areas with medium rainfall and light snowfall. A *high gable* sheds rain and snow better and affords more loft and storage space. A *gambrel* ("Dutch barn") shape provides more headroom and stor-

age space than a gable. The *"saltbox"* is usually found on America's east coast: a steep gable with a shed extension on one side. The *hip* roof is often built in areas of high rainfall; the four sloping sides of the roof protect all walls from falling rain.

Choosing a roof shape involves many factors, some covered in the preceding pages on design considerations: study of climate and site, practical use of the building, analysis of other buildings in the area, energy considerations, etc.

On the following 14 pages are framing drawings (by Bob Easton) and photos of seven basic roof shapes: the most common shapes used throughout the world for rectangular buildings. A small floor plan is shown with each shape; the floor plan is keyed to the roof framing, showing how floor plan and framing (as well as foundation) are interrelated.

The drawings are intended as visual guides, not as plans to build from. Local conditions such as wind force, snow loads, earthquake potential, dampness, etc. should be carefully considered before choosing a roof shape, and will be instrumental in adapting these shapes to your particular site and needs. We suggest consulting the local building inspector as to the structural requirements of your area. Even if a permit is not required, the codes will most likely be adapted to local conditions and will tell you things these drawings do not. And in most cases, the building inspector will be willing to offer advice. It would also be wise to talk to a local builder.

The roof shapes shown are not only for stud frame walls. They can be adapted to other types of wall construction, such as adobe, stone, or masonry.

Flat Roof

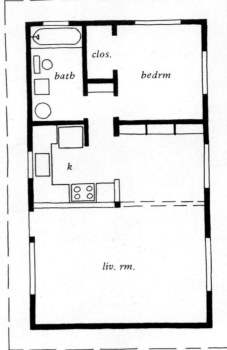

Floor plan

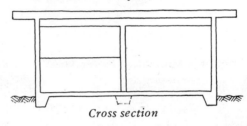

Cross section

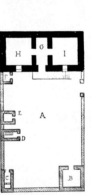

Buildings with flat roofs are generally found in areas with light rainfall and moderate snowfall, such as the desert. Adobe buildings are typically built with flat roofs, and the shape is often used where people spend much of their time outside, in patios, gardens or courts. A flat roof is probably the easiest shape to add on to.

A flat roof can be the first stage in construction of a two-story house, with the following conditions:
— foundations for a two-story building must be poured.
— roof rafters should be floor joist-size.
— roof must be constructed level, with no slope for water run-off.

For water run-off, pitch roof minimum 1" per 10'. (Can be done by adding 1 or 2 2x4's to one plate.) If not pitched, roof will sag, water will puddle in middle.

Consult local codes for lumber species, grade, and allowable spans.

Add footings under wall betw. liv. rm. and bedroom and under post.

Adobe house

Floor plan and cross section at 3/32" = 1'-0" scale. This scale is on triangular rulers available at drafting supply stores.

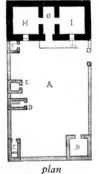

plan

A—small garden
B—pantry
C—privy
D—dovecotes
E—chicken house
F—oven

Egyptian rural dwelling

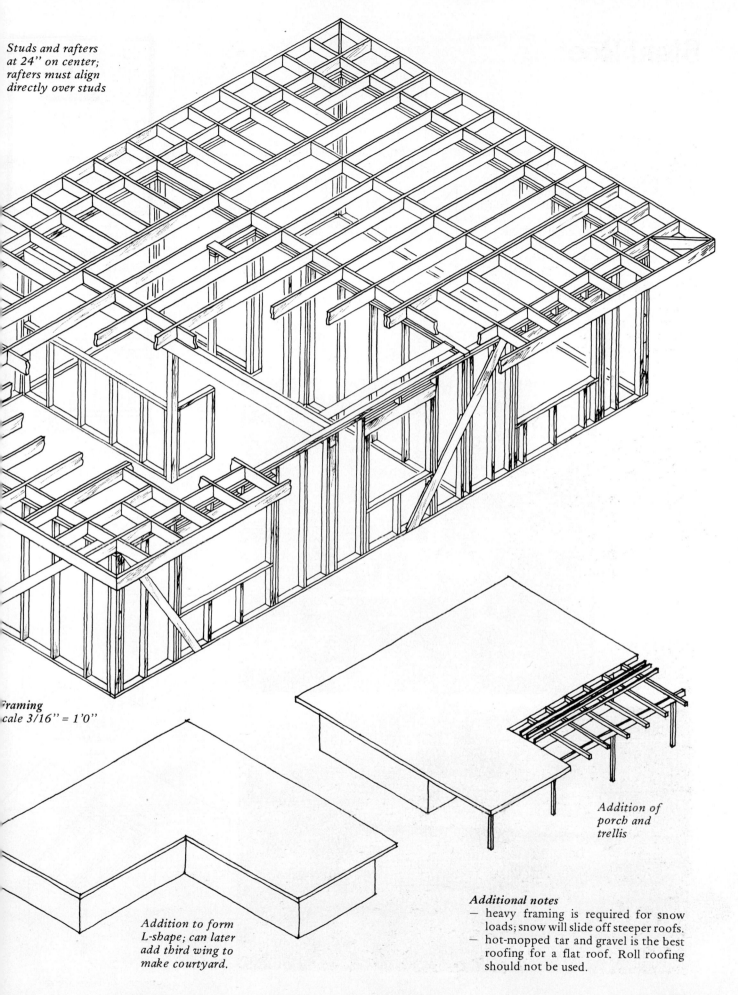

*Studs and rafters
at 24" on center;
rafters must align
directly over studs*

Framing
cale 3/16" = 1'0"

*Addition to form
L-shape; can later
add third wing to
make courtyard.*

*Addition of
porch and
trellis*

Additional notes
— heavy framing is required for snow
  loads; snow will slide off steeper roofs.
— hot-mopped tar and gravel is the best
  roofing for a flat roof. Roll roofing
  should not be used.

Shed Roof

A shed roof is a simple shape to build, sheds water and snow better than a flat roof, and is a good shape for later additions or extensions. It is also a good shape for adding to an existing building. Clerestory (high) windows are often installed on the high side of a shed roof building, allowing high light to enter without the waterproofing problems of skylights.

Shown here is a small shed building with a six foot wide loft. The smaller drawings show additions to the shed shape. When building overhang on shed, nail securely with 4-16d the rafters to top plate.

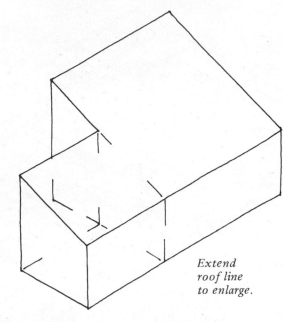

Extend roof line to enlarge.

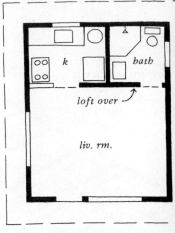

Floor plan
Scale: 3/32" = 1'-0"

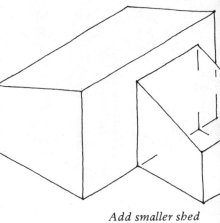

Cross section

Add smaller shed to high side.

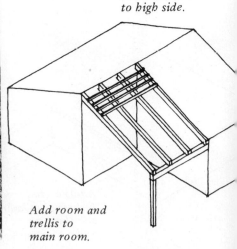

Add room and trellis to main room.

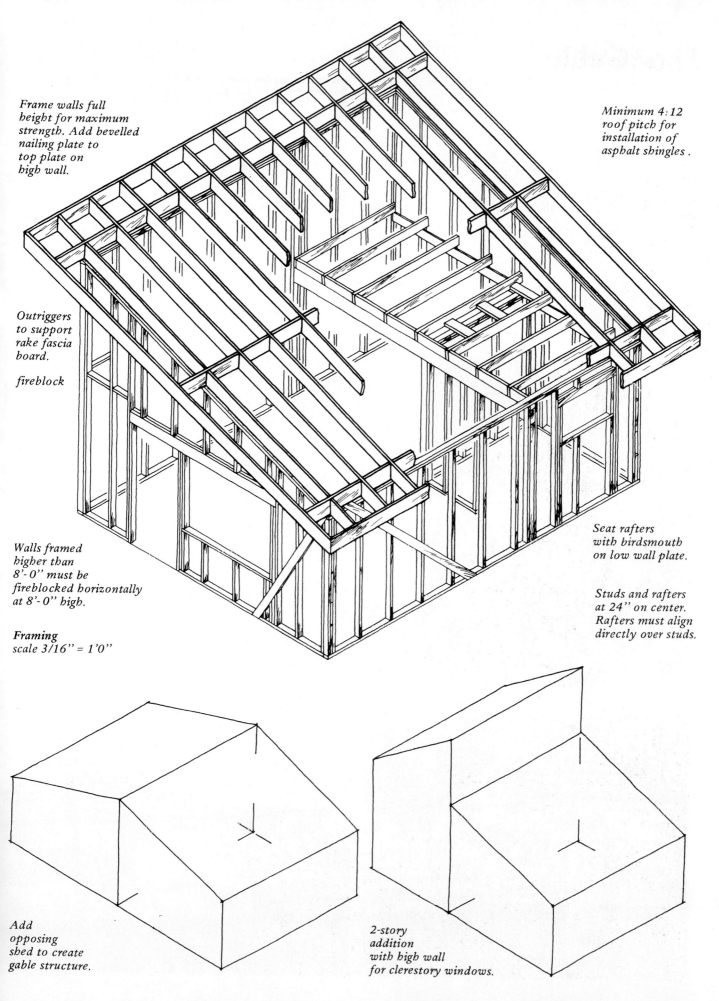

Frame walls full
height for maximum
strength. Add bevelled
nailing plate to
top plate on
high wall.

Minimum 4:12
roof pitch for
installation of
asphalt shingles .

Outriggers
to support
rake fascia
board.

fireblock

Walls framed
higher than
8'- 0" must be
fireblocked horizontally
at 8'- 0" high.

Framing
scale 3/16" = 1'0"

Seat rafters
with birdsmouth
on low wall plate.

Studs and rafters
at 24" on center.
Rafters must align
directly over studs.

Add
opposing
shed to create
gable structure.

2-story
addition
with high wall
for clerestory windows.

Low Gable

The low gable is probably the most common roof in America. It has the advantage of using shorter, thus smaller sized, roof rafters than either a flat or shed roof. If secured properly with ceiling joists or cross ties, it has the structural stability of a triangle.

It can have flat ceilings or open, or a combination: flat over partitioned rooms (bathroom, bedroom, etc.), open over living room.

Use trusses for longer spans than shown here. Building codes require engineering for spans over 24 feet.

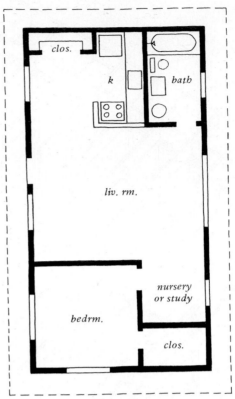

Floor plan
scale: 3/32" = 1'-0"

Pour footing under wall between bath and kitchen, will function as bearing wall.

Install 2 rows horiz. blocking if vertical siding is used.

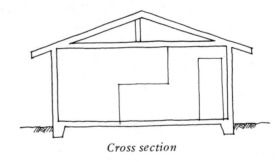

Cross section

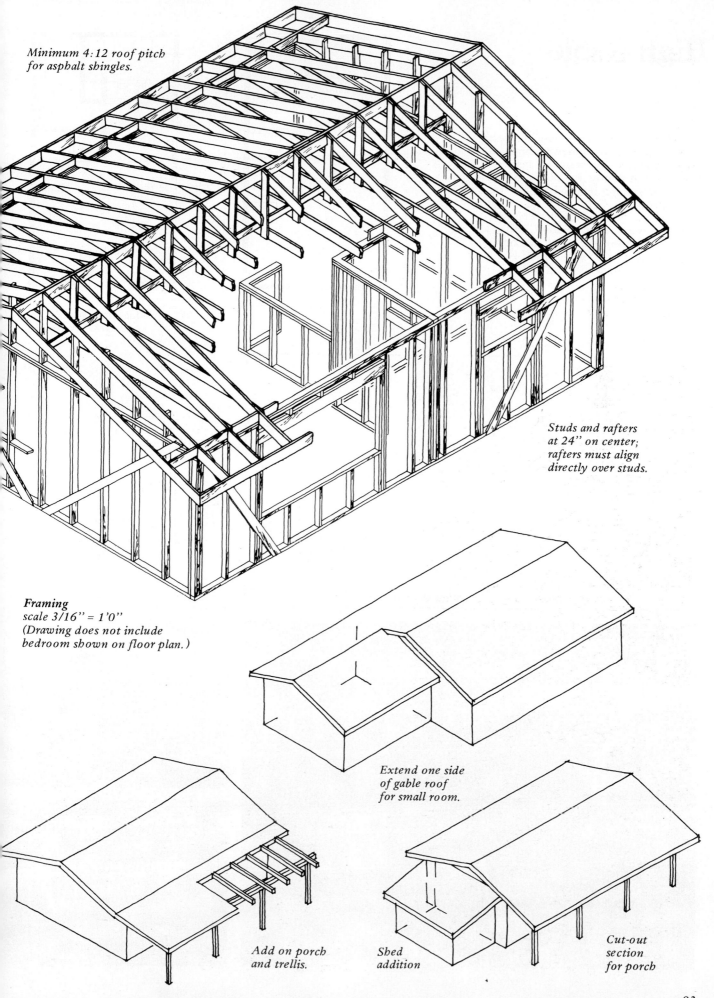

Minimum 4:12 roof pitch
for asphalt shingles.

Studs and rafters
at 24'' on center;
rafters must align
directly over studs.

Framing
scale 3/16'' = 1'0''
(Drawing does not include
bedroom shown on floor plan.)

Extend one side
of gable roof
for small room.

Add on porch
and trellis.

Shed
addition

Cut-out
section
for porch

High Gable

A steep gable roof is often used in areas with moderate to heavy rainfall or heavy snowfall. The steepness helps to shed water and snow and allows enough space for storage or a loft above the plate level.

Although the framing drawing here shows an open ceiling and loft (framed with ridge beam), the more traditional high gable is framed with a ridge board and joists at plate level (see p. 93).

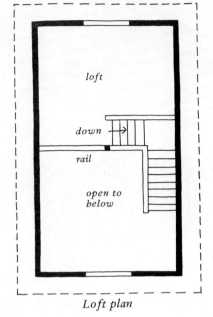

Loft plan

Main floor plan
scale: 3/32" = 1'- 0"

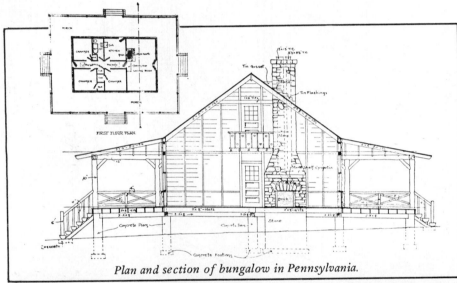

Plan and section of bungalow in Pennsylvania.

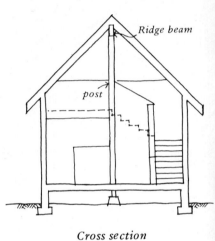

Cross section

94

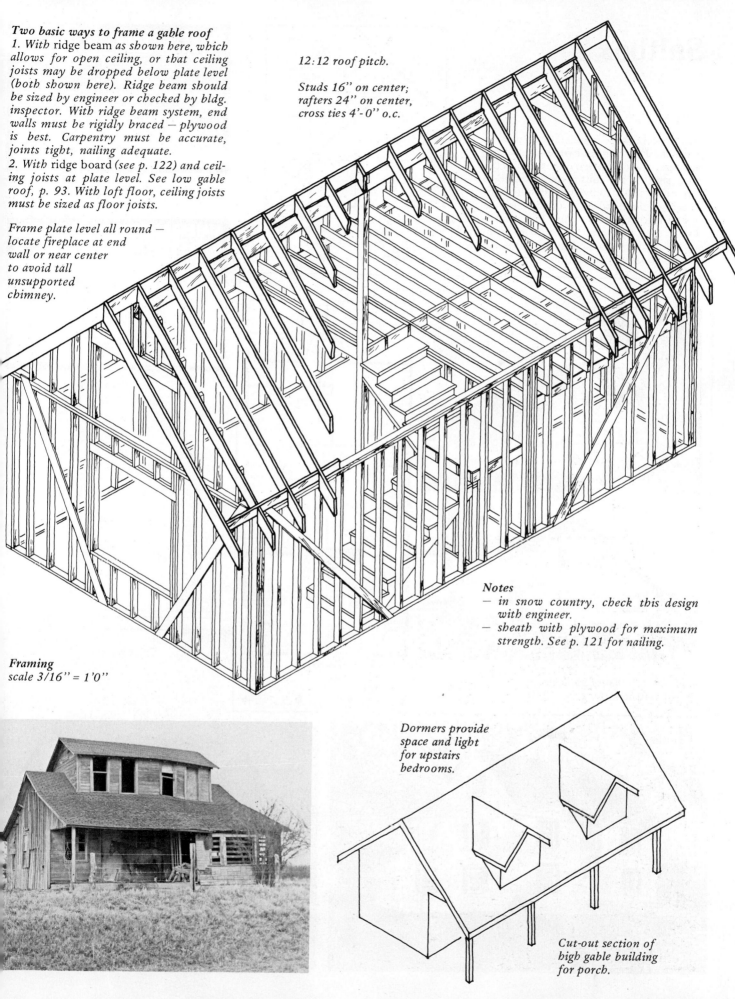

Two basic ways to frame a gable roof
1. With ridge beam *as shown here, which allows for open ceiling, or that ceiling joists may be dropped below plate level (both shown here). Ridge beam should be sized by engineer or checked by bldg. inspector. With ridge beam system, end walls must be rigidly braced — plywood is best. Carpentry must be accurate, joints tight, nailing adequate.*
2. With ridge board *(see* p. 122*) and ceiling joists at plate level. See low gable roof,* p. 93. *With loft floor, ceiling joists must be sized as floor joists.*

Frame plate level all round — locate fireplace at end wall or near center to avoid tall unsupported chimney.

12:12 roof pitch.

Studs 16" on center; rafters 24" on center, cross ties 4'-0" o.c.

Notes
— *in snow country, check this design with engineer.*
— *sheath with plywood for maximum strength. See* p. 121 *for nailing.*

Framing
scale 3/16" = 1'0"

Dormers provide space and light for upstairs bedrooms.

Cut-out section of high gable building for porch.

95

Saltbox

The saltbox shape is generally associated with the New England states and severe winters. They are often oriented with the high side to the south, the low side to the north. This allows winter sun to hit the high side, and snow (a good insulator) to accumulate on the lower, shallower roof to the north. Snow or bales of hay are often banked against the north side in winter for insulation.

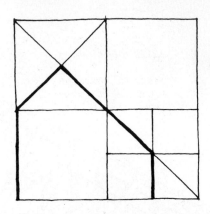

Classic New England saltbox profile, ca. 1690. After drawing by Eric Sloane.

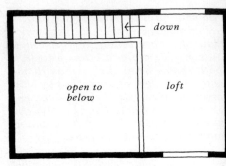

Loft floor plan

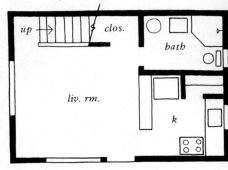

Main floor plan
Scale: 3/32 = 1'-0"

Saltbox farmhouse published in The Craftsman, *January, 1909.*

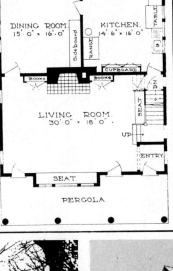

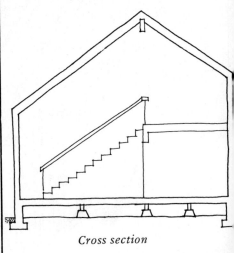

Cross section

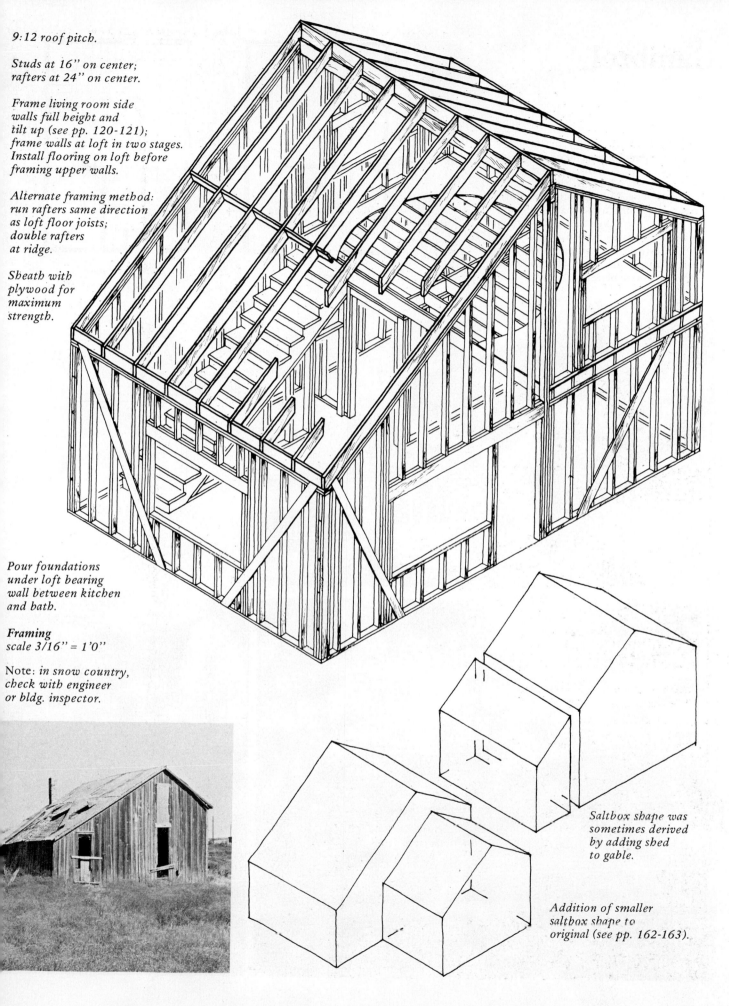

9:12 roof pitch.

Studs at 16" on center; rafters at 24" on center.

Frame living room side walls full height and tilt up (see pp. 120-121); frame walls at loft in two stages. Install flooring on loft before framing upper walls.

Alternate framing method: run rafters same direction as loft floor joists; double rafters at ridge.

Sheath with plywood for maximum strength.

Pour foundations under loft bearing wall between kitchen and bath.

Framing
scale 3/16" = 1'0"

Note: *in snow country, check with engineer or bldg. inspector.*

Saltbox shape was sometimes derived by adding shed to gable.

Addition of smaller saltbox shape to original (see pp. 162-163).

Gambrel

Gambrel roofs are most often found in the eastern part of the United States and Canada. The word derives from the hock (bent part) of a horse's leg, also called a gambrel. The lower part of the roof is a steep slope, the upper part shallower. The break in roof line allows head room in the loft space, and is useful in barns for hay storage (see pages 102-103 for gambrel barn plans), as well as in homes for rooms above plate level.

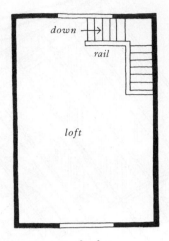

Loft plan

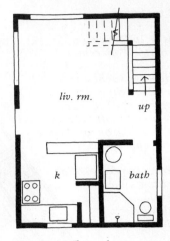

*Main floor plan
scale: 3/32" = 1'-0"*

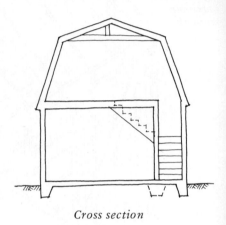

Cross section

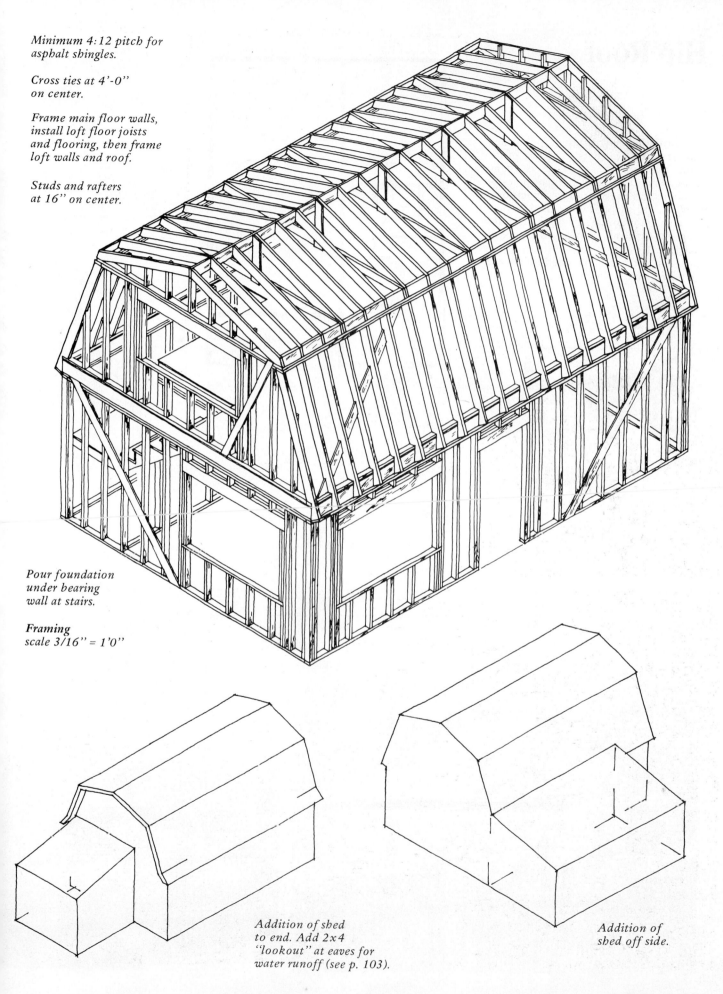

Minimum 4:12 pitch for
asphalt shingles.

Cross ties at 4'-0"
on center.

Frame main floor walls,
install loft floor joists
and flooring, then frame
loft walls and roof.

Studs and rafters
at 16" on center.

Pour foundation
under bearing
wall at stairs.

Framing
scale 3/16" = 1'0"

Addition of shed
to end. Add 2x4
"lookout" at eaves for
water runoff (see p. 103).

Addition of
shed off side.

Hip Roof

A hip roof has four different roof planes. It is an attractive shape, and one of the most complex to frame. It is found throughout the world, built of poles or bamboo, and in America of dimensional lumber. The shape is often used in areas of high rainfall. With no high walls (above plate level) there is less chance of leakage. Because of difficulties in framing, it is not a recommended roof shape for a first-time builder.

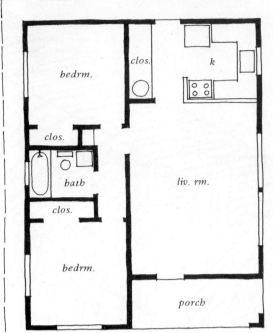

*Floor plan
scale: 3/32" = 1'-0"*

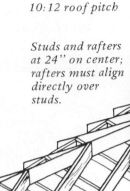

10:12 roof pitch

Studs and rafters at 24" on center; rafters must align directly over studs.

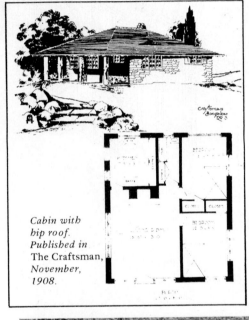

Cabin with hip roof. Published in The Craftsman, November, 1908.

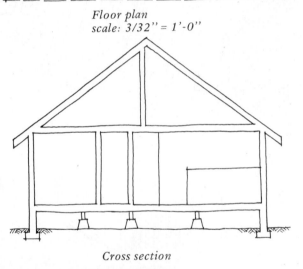

Cross section

Frame walls same height all around (incl. bearing wall).

Pour foundation under bearing wall between liv. rm., kitchen and bedrooms.

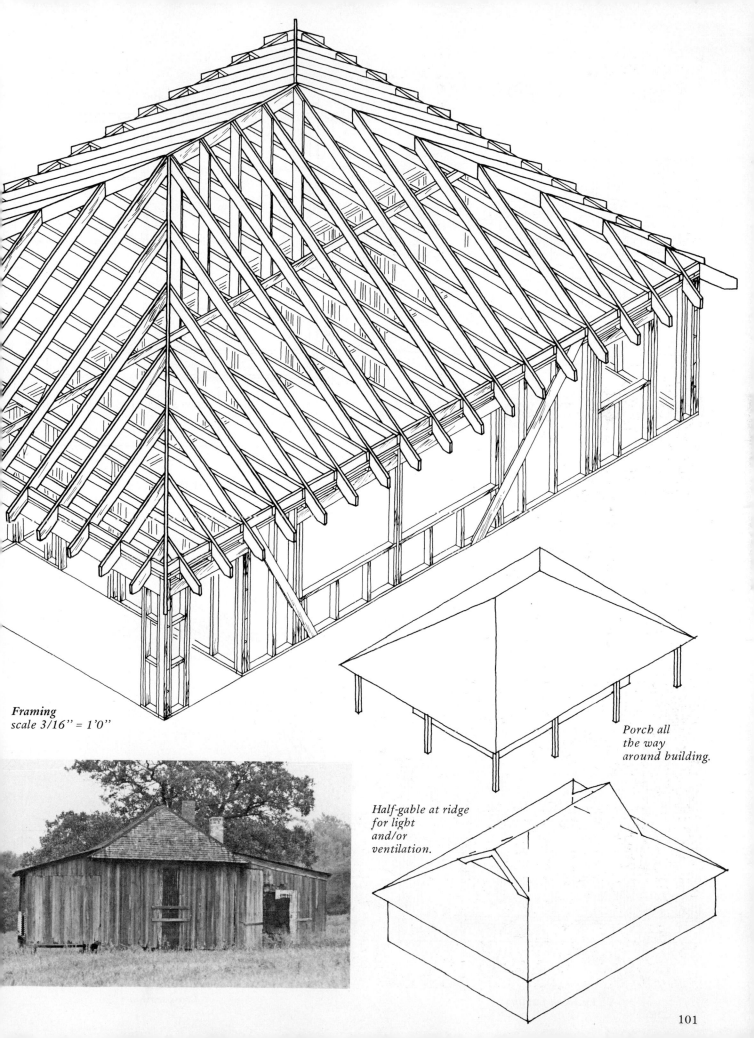

Framing
scale 3/16" = 1'0"

*Porch all
the way
around building.*

*Half-gable at ridge
for light
and/or
ventilation.*

101

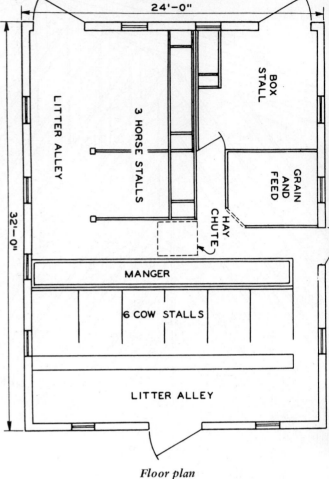

Floor plan

Gambrel Barn

These plans are for a 24 by 32 foot gambrel-roof barn of simple construction. Short lengths of lumber can be used and no large or heavy timbers are required. The haymow of this barn has a capacity of 15 tons of loose hay. Drawings from *Fundamentals of Carpentry - Volume 2, Third Edition,* by Walter E. Durbahn & Elmer W. Sundberg. Reprinted by permission of American Technical Society.

End-wall framing

Cross section

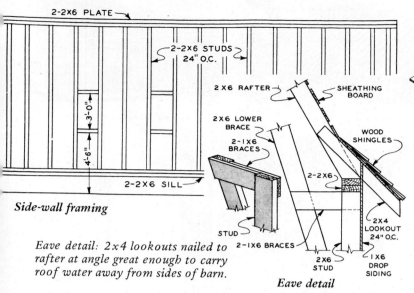

Side-wall framing

2-2X6 PLATE

2-2X6 STUDS 24" O.C.

3'-0"

4'-6"

2-2X6 SILL

Eave detail: 2x4 lookouts nailed to rafter at angle great enough to carry roof water away from sides of barn.

2 X 6 RAFTER

SHEATHING BOARD

2 X 6 LOWER BRACE

2-1X6 BRACES

WOOD SHINGLES

2-2X6

STUD

2-1X6 BRACES

2X6 STUD

2X4 LOOKOUT 24" O.C.

1X6 DROP SIDING

Eave detail

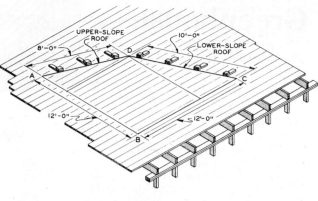

UPPER-SLOPE ROOF

8'-0"

10'-0"

LOWER-SLOPE ROOF

A D C

12'-0" 12'-0"

B

Above: Rafter bents are laid out on floor of haymow. Point A = peak of roof. Note that the 8'-0" and 10'-0" rafter lengths are also shown on cross section. The upper and lower rafter bents are held in place on floor with 2x4 blocks. Note how each rafter forms the third side of a right triangle.

Backyard Hen House

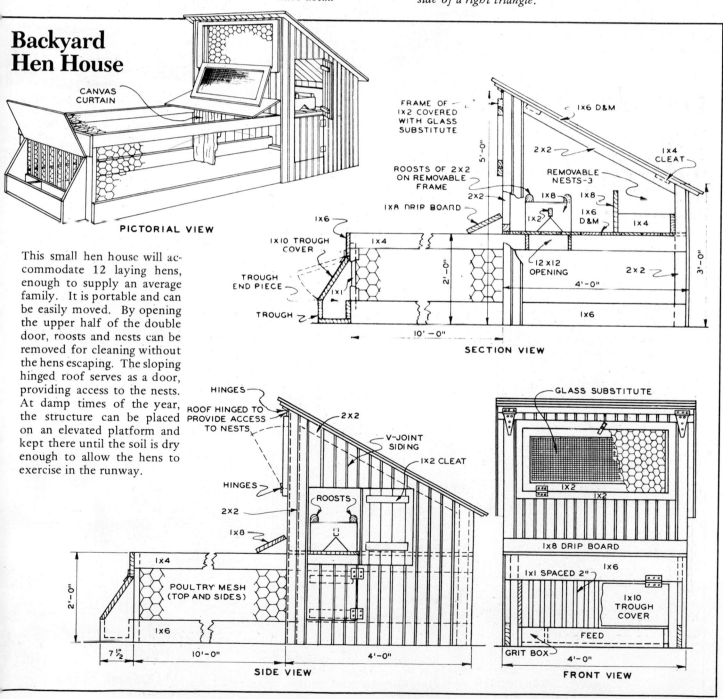

CANVAS CURTAIN

PICTORIAL VIEW

This small hen house will accommodate 12 laying hens, enough to supply an average family. It is portable and can be easily moved. By opening the upper half of the double door, roosts and nests can be removed for cleaning without the hens escaping. The sloping hinged roof serves as a door, providing access to the nests. At damp times of the year, the structure can be placed on an elevated platform and kept there until the soil is dry enough to allow the hens to exercise in the runway.

FRAME OF 1x2 COVERED WITH GLASS SUBSTITUTE

1x6 D&M

5'-0"

2x2

1x4 CLEAT

ROOSTS OF 2x2 ON REMOVABLE FRAME

REMOVABLE NESTS-3

2x2

1X8 DRIP BOARD

1x8   1x8

1x2   1x6 D&M

1x4

1x6

1x4

1x10 TROUGH COVER

2'-0"

12 x 12 OPENING

4'-0"

2x2

3'-0"

TROUGH END PIECE

1x1

TROUGH

1x6

10'-0"

SECTION VIEW

HINGES

ROOF HINGED TO PROVIDE ACCESS TO NESTS

2x2

V-JOINT SIDING

1x2 CLEAT

GLASS SUBSTITUTE

HINGES

2x2

1x8

ROOSTS

1x4

POULTRY MESH (TOP AND SIDES)

2'-0"

1x6

1x8 DRIP BOARD

1x1 SPACED 2"

1x6

1x10 TROUGH COVER

FEED

GRIT BOX

7 1/2"

10'-0"

4'-0"

SIDE VIEW

4'-0"

FRONT VIEW

Greenhouses

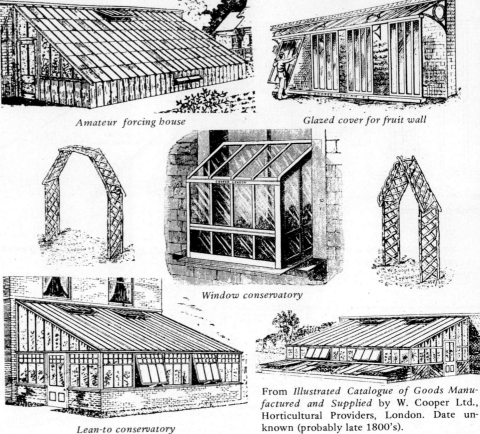

Amateur forcing house

Glazed cover for fruit wall

Window conservatory

Lean-to conservatory

From *Illustrated Catalogue of Goods Manufactured and Supplied* by W. Cooper Ltd., Horticultural Providers, London. Date unknown (probably late 1800's).

Points to consider in greenhouse construction

– shade can be provided in summer by a row of deciduous trees planted in front, or vines (like grapes) growing inside.

– thermal mass (adobe north wall, drums of water, etc.) retains heat after sun goes down.

– ventilation is critical. Vents in roof and low on walls create up-draft. Be sure to have roof vents within easy reach due to frequency of use.

– bug infestation can be minimized by keeping greenhouse clean, rotating crops, introducing pest eaters such as ladybugs or praying mantises.

– north wall can be opaque.

– coldframes are much smaller and simpler to build than greenhouses and can be used for starting plants.

Cold Frame/Greenhouse

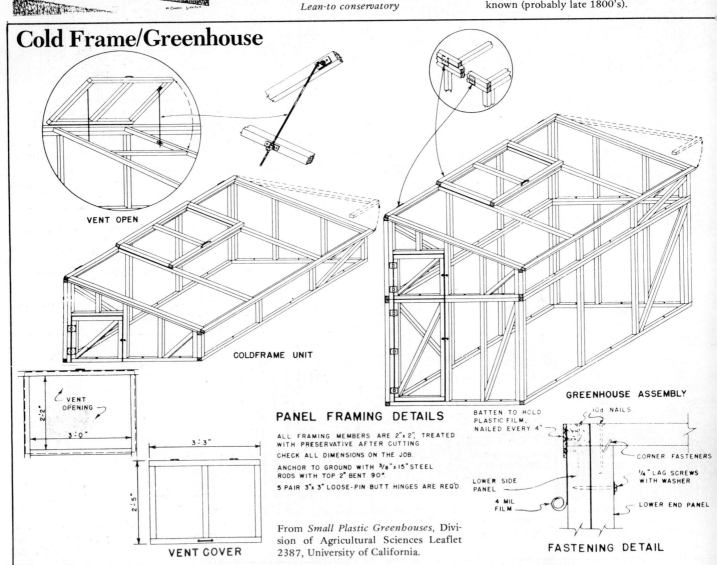

VENT OPEN

COLDFRAME UNIT

VENT OPENING
2'-2"
3'-0"

VENT COVER
3'-3"
2'-5"

PANEL FRAMING DETAILS

ALL FRAMING MEMBERS ARE 2"x 2", TREATED WITH PRESERVATIVE AFTER CUTTING.

CHECK ALL DIMENSIONS ON THE JOB.

ANCHOR TO GROUND WITH 3/8"x15" STEEL RODS WITH TOP 2" BENT 90°.

5 PAIR 3"x 3" LOOSE-PIN BUTT HINGES ARE REQ'D.

From *Small Plastic Greenhouses*, Division of Agricultural Sciences Leaflet 2387, University of California.

GREENHOUSE ASSEMBLY

BATTEN TO HOLD PLASTIC FILM, NAILED EVERY 4"

10d NAILS

CORNER FASTENERS

1/4" LAG SCREWS WITH WASHER

LOWER END PANEL

LOWER SIDE PANEL

4 MIL FILM

FASTENING DETAIL

The Building Codes

by Robert Thallon

Every man who builds his own house is naturally interested in building a house that will provide for the health, safety, and comfort of himself and his family. These are instinctive concerns which have been shared by all owner-builders throughout history. It is only when a house is built by someone other than the owner himself — by a professional builder — that this natural interest in building a healthful, safe, and comfortable house can be lost. The professional builder may not share the owner-builder's keen interest in these matters because he will not be living in the house he builds; he may simply be building the house to earn money. (However, a conscientious builder — mindful of his reputation — will strive to produce the highest quality house that his budget will allow.)

In order to insure that professional builders build good houses, it has become necessary to impose minimum standards or codes that must be followed. The earliest known building code is contained in the code of Hammurabi, the ruler of Babylon in the 18th century B.C. Hammurabi's code dealt with the strength of buildings, making the professional builder responsible for his own work. If a house fell down killing its occupant, the builder of the house was to be slain.

Early U. S. building codes

The first building code in this country was enacted in Plymouth Colony in the year 1629. This code was aimed at lowering the risk of fire by prohibiting the construction of thatched roofs in the city. Other colonies followed this example, and many codes aimed at lowering fire risk were adopted; but it was not until over 200 years later, in the year 1850, that the first building code relating to health was proposed. In that year the Sanitary Commission of Massachusetts recommended that local Boards of Health "... endeavor to prevent or mitigate the sanitary evils arising from over-crowded lodging houses and cellar-dwellings ..." Seventeen years later (1867) these recommendations were embodied in the first Tenement House Act of the State of New York. This law provided for the health and safety of tenement and lodging-house dwellers by requiring that: (1) sleeping rooms be ventilated; (2) a water closet be provided for every twenty occupants; (3) the dwelling be kept clean to the satisfaction of the Board of Health; and (4) stairways be equipped with banisters. The concern for the health and safety of dwellers expressed in this, the first of the American building codes, is still embodied in the building codes of today. The preface to the currently most widely used building code states that "The standards of safe construction that are set out in the Uniform Building Code are primarily founded upon health and safety ..."

Public health & safety

Protection of the public health and safety from unscrupulous and unknowledgeable building practices is a noble objective, and one that has become increasingly important in recent years. Improved transportation and rapid population growth have put people (and houses) in such close proximity to each other that public health regulations are more vital today than ever before. Hardly a building site remains where improperly treated sewage wouldn't affect a neighbor's water supply. The need for regulations of public safety has also intensified as electricity and its accompanying technological advances have increased the danger of fire and introduced a new danger, electrical shock. An awareness of the increased need for regulations to protect the public from these and other hazards has resulted in four separate classes of codes which apply to building a house:

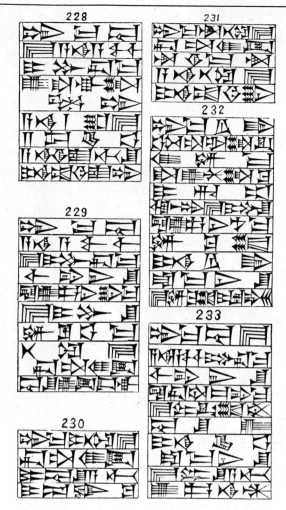

The Building Code of Hammurabi, Founder of the Babylonian Empire; earliest known code of law; translated:

228. If a builder build a house for a man and complete it, that man shall pay him two shekels of silver per sar (approximately 12 square feet) of house as his wage.

229. If a builder has built a house for a man and his work is not strong, and if the house he has built falls in and kills the householder, that builder shall be slain.

230. If the child of the householder be killed, the child of that builder shall be slain.

231. If the slave of the householder be killed, he shall give slave for slave to the householder.

232. If goods have been destroyed, he shall replace all that has been destroyed; and because the house that he built was not made strong, and it has fallen in, he shall restore the fallen house out of his own material.

233. If a builder has built a house for a man, and his work is not done properly and a wall shifts, then that builder shall make that wall good with his own silver.

— *Building Codes:* There are four building codes used in this country, but all four are almost identical in their major requirements.

Sanitary Codes: These locally controlled codes are more widespread than the building codes. As applied to house building, they are generally concerned with insuring adequate disposal of human waste ...

— *Electric Codes:* The national Electric Code, written by electrical inspectors, underwriters, power company employees, and electrical equipment manufacturers is virtually a national code. Power companies will not connect electricity to a house unless there is proof of compliance with the National Electric Code.

— *Plumbing Codes:* There is a national Plumbing Code but it is usually only enforced in areas requiring a building permit ... *continued*

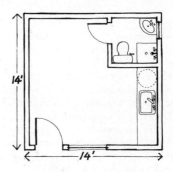

Sec. 1405. (a) Light and Ventilation. *All . . . habitable rooms within a dwelling unit shall be provided with natural light by means of windows or skylights with an area of not less than one-tenth of the floor area of such rooms. All bathrooms . . . shall be provided with natural ventilation by means of windows or skylights with an area of a minimum of 3 square feet.*

Not less than one-half of the required window or skylight area shall be openable to provide natural ventilation.

(b) Sanitation. *A room in which a water closet is located shall be separated from food preparation or storage rooms by a tight-fitting door.*

Every dwelling unit shall be provided with a kitchen equipped with a kitchen sink and with bathroom facilities consisting of a water closet, lavatory and either a bathtub or shower. Plumbing fixtures shall be provided with running water necessary for their operation.

Sec. 1407. (a) Ceiling Heights. *Habitable rooms . . . shall have a ceiling height of not less than 7 feet 6 inches.*

(b) Superficial Floor Area. *Every dwelling unit shall have at least one room which shall have not less than 120 square feet of superficial floor area. Every room which is used for both cooking and living or both living and sleeping purposes shall have not less than 150 square feet of superficial floor area . . . exclusive of fixed or built-in cabinets or appliances.*

Sec. 1410. Every dwelling unit and guest room shall be provided with heating facilities capable of maintaining a room temperature of 70° F at a point 3 feet above the floor in all habitable rooms.

If these codes are to be effective in their function as protectors of the public health and safety, every house must come under their jurisdiction. Even owner-built houses, whose owners fully intend to build a safe and healthful house, must be subject to the regulations of the codes because, despite their good intentions, owner-builders are often uninformed. Just because an owner-builder wants to build a safe chimney doesn't mean that he will be able to. If an owner-built house catches fire because of a faulty chimney or faulty electrical system, firemen are called to risk their lives just the same as for any other fire. Neighboring houses and adjoining forests catch fire just as readily from a flaming owner-built house as from a professionally-built house. Public streams and waterways can be contaminated from faulty owner-built sewage systems as easily as from any faulty sewage system. And even if an owner-builder contaminates only his own water system, his resulting sickness becomes a public problem. The need to treat owner-builders the same as any builder in regard to the protection of the public health and safety should be clear from these examples. The argument is even more convincing, however, when we consider that there can be no guarantee that the owner-builder will not sell his house, thereby putting himself in the same position as the professional builder for whom the codes were originally intended.

Despite the apparent soundness of this reasoning, a large percentage of owner-builders resent the restrictions of the codes. These people want nothing to do with the codes, and they will go to almost any extreme to avoid being governed by them. The reasons given for their reluctance to be governed by the codes go beyond the money saved on the cost of a permit to the codes themselves, whose regulations tend to extend beyond the areas of health and safety.

The following illustration shows a plan of the smallest, most minimal house allowed by the Uniform Building Code. It has the least floor space and as few conveniences as the code will allow. Although I am sure no one would ever want to, this tiny house could be built almost anywhere . . .

Although most people would prefer more than the minimal conveniences required by the code, some of these requirements clearly extend beyond the realm of health and safety into the area of comfort. We can be sure that the requirement for a water closet (toilet), for example, is a question of comfort and not of health since latrines (outhouses) can be found in practically every state and national park. Of course, latrines should not be allowed in heavily populated areas, but this is no reason to deny everyone the choice between a latrine and a water closet. I have talked with many owner-builders who claim to be quite comfortable using a latrine. They have applied the money which would have been necessary to install a water closet and septic system toward some other comfort.

Codes go beyond health and safety

It is easy to understand why the codes have gone beyond health and safety into the area of comfort when we recall that they were originally enacted in order to protect the tenement and lodging-house dwellers of the nineteenth century from the outrageous practices of the landlords and slum-builders of that period, and when we recognize that they apply to slums and slum-builders of the present day. But owner-builders are keenly interested in their own comfort and are the ones who know best what they will find comfortable. There is no need to regulate comfort for owner-builders. The codes, by attempting to do this, are only working to defeat their own purpose of protecting the public health and safety by alienating owner-builders who might otherwise not connive at avoiding the codes altogether.

Another complaint I have heard from owner-builders is that the code must be strong enough to include the very worst solution within the set of acceptable solutions. In order to allow for the worst design and sloppiest craftsmanship, the code must set minimum standards above the standards that would be acceptable with good design or good craftsmanship. As a result, good work doesn't count; good design must still meet the requirements of bad design. The code requires, for example, that each habitable room have 90 square feet of superficial floor area with no dimension less than 7 feet. If the smallest room allowed by the code (7 feet by 13 feet) were used for a bedroom, we would have trouble fitting in the furniture. A bedroom with a built-in bed and built-in storage and seating, however, could be easily designed with only 15 square feet of superficial floor area (see fig. 49). But the code requires 90 square feet! Regulations like this tend to irritate owner-builders who, for reasons of economy, often strive to build their houses as small as possible.

Codes require permanence

It is the intent of the code to require a method of construction that is permanent. The preface to the Uniform Building Code, Vol. VI, states that "The Uniform Building Code . . . is comprehensive and flexible so that no material or method of construction that is permanent and safe is excluded." Again, it is easy to see the reasoning behind this requirement when we think of the terribly delapidated slum housing existing today in the cities, but this purpose is in conflict with the desires of those owner-builders who want to build and inhabit a house temporarily. Sometimes owner-builders want to live in a temporary house while designing and building a more permanent dwelling. Sometimes they have plans for moving on. Whatever the reason they have for choosing to build a temporary house, the code regulations requiring permanent construction seems quite arbitrary to them. In these times of rapid change, it is possible that what is needed is less permanence and more temporary experimentation. Can you imagine what the land would look like now if all the Indians had been required to put a continuous concrete foundation under their houses . . . ?

Throughout this article I have been arguing that the codes ought to restrict the owner-builder only when his actions might endanger the public health or safety. I argue that the owner-builder deserves special consideration under the codes since, unlike the professional builder, he is naturally interested in the comfort of the future occupants of the house (his own family), and he is therefore naturally interested in the quality of his house. He is in the special position of being able to directly translate his own needs, desires, and capabilities into the making of his house (given that he is a competent builder and selects good quality materials). Even though the codes themselves do not recognize this special relationship, owner-builders who choose to follow the codes usually find special considerations given them indirectly through the enforcement of the codes.

It is clear that, in order to obtain a building permit, the building department needs to be convinced that the regulations of the code will be followed in every detail. Once the permit has been issued, however, the owner-builder finds that he need not follow the code as closely as he said he would. He finds that the building inspector is a different person in the field than in the office. Inspectors in the field are free to use their own discretion in matters of compliance with the code. They are at once the interpreters and enforcers of the code. This is where the owner-builder will receive special consideration. It is refreshing for an inspector, who spends most of his day inspecting industrial building and tract houses, to see an owner-built house. He can appreciate the effort involved in building a house, and realizes that owner-builders are not accustomed to building techniques nor are they accustomed to the regulations of the code. Consequently, the inspector will frequently make allowances for this that he would never make to the professional builder. These allowances are usually slight; the codes are stretched rather than ignored. You don't see inspectors waiving the requirements for a bathroom, but rather extending the distance specified in the code. These concessions are almost always made after the work has been completed, because it is more difficult for an inspector to tell an owner-builder to change his work than to tell him to change his plans. The owner-builder has cause to be grateful for the special consideration he receives, but it is regrettable that this consideration is not more far-reaching and that it must come, as it does, through the back door via the personal judgement of each inspector.

In a sense it is quite unfortunate that most experimental house builders have chosen to evade the building codes. By evading the codes they also evade the fundamental question of whether a person has a right to build a house for himself in any way he sees fit as long as he doesn't endanger the public health or safety . . . . □

Building Inspector

Herb Wimmer began his building career at age 12, working as a lather for his father in St. Louis. He then worked for a small contractor during the depression (the '30's), three years with his uncle building houses and barns in Missouri, and another few years on heavy construction in New York. From 1945-1962 he was an independent contractor in the San Francisco Bay Area. From 1962 until 1978 he has worked in three different building inspection departments, the last eight years as Marin County, California's Chief Building Inspector.

L.K.: *How long do you think a building should last?*

Herb Wimmer: Well, this question's been asked before and facetiously you could say it should last at least as long as the mortgage. But, quite frankly, I think that wood frame buildings that are built adequately will probably have a life expectancy of several hundred years. Masonry construction, again built adequately, can have a life span of half a century or more. Now, obviously when I say a couple of hundred years, it has to be maintained. Certainly the roofing is not going to last two hundred years; the average roofing lasts twenty to thirty years . . . . I've known buildings that are a hundred years old and have never been maintained; most of them are in pretty sad shape by that time; had they been maintained, they'd probably have lasted another hundred years.

Does the Uniform Building Code apply in most of the United States?

Oh, no. We have the Southern Building Code Conference, which basically is used in Florida, South Carolina, Alabama — in the southeastern region. Then, west of that we have the Basic Building Code, published by the Building Officials Conference of America, Inc., called *BOCA*; that generally takes over in the central part of the United States, east of the Mississippi River, although the Uniform Building Code is also used in some areas east of the Mississippi. And the Uniform Building Code is used in the western half of the United States.

Do the counties decide if they want to adopt the major codes?

The counties or the cities decide if they want to write their own code and if they do not, if they don't have the expertise, then they choose the model code that suits their circumstances. Many of the large cities still write their own codes. The city of New York is a good example; they have their own testing centers and their own research facilities . . . .

What would you say were originally the most important reasons for building codes?

Originally, the building codes started out, not so much for health considerations as for fire and safety considerations.

Fire and safety? What about structural safety?

I don't think structural safety entered

into it, because when America was a young country there were no highrise buildings.

They weren't worried about buildings falling down?

They were mostly concerned with fire. In fact, all the early regulations dealt with fire — dealt with such things as the elimination of thatched roofs, the elimination of loft chimneys, the abatement of abandoned buildings which were causes of fire. So that the original considerations in this country dealt with fire as it related to construction.

And then health?

Yes. The health aspect came much later, though. It came after some severe health epidemics in the East. Of course, that was before this part of the country was even inhabited.

And as the years have gone by, they've actually added more and more to the codes.

Rather than saying "added to it" I would say they've filed all the regulations in one document rather than in a great number of documents or individual ordinances, regulations that were scattered and the responsibilities of a large number of departments or agencies. By combining them into one document, the basic information was carried in one place rather than in a number of separate places.

Someone building a house without much experience . . . what do you think are the most important aspects of the code as far as just helping that person building the house and members of the public who would be either living in the house afterwards or neighbors?

Of course we don't make any distinctions whether it's owner-occupied or rented. We're still talking basically about life safety from the standpoint of fire safety and structural safety, as well as health safety . . . .

What do you think about drawing up plans? Do you have any thoughts about what is acceptable to a building department as far as the submission of plans by an owner-builder?

Well, this of course varies from one building to another. We need to know, of course, the size of the structure, that it has the proper kind of footings, the proper framing, the proper light and ventilation. We will accept plans drawn on almost anything so long as the plans tell us yes, this building is going to comply with the requirements of the code. It does not have to be done by an architect; it does not have to be done by an engineer. When homeowners, however, start playing around with spans that are in excess of 25 feet, and some of them do, then we get into a requirement for engineering.

With say, a post and beam house, do you require an engineer to look it over for shear panels, things like that?

Well, that depends on the amount of openings that people have. It's relatively easy to determine whether they're going to have enough shear resistance by just looking at the plan. If there appear to be problems then we might ask them to have an engineer check.

What do you think of a post and beam house as compared to a stud house? The reason I ask is that I built a post and beam house one time: I got it all framed and then I said, I'm going to put my walls on, and what am I going to put in between the posts which are eight feet apart? Well, I ended up putting a stud wall in between the posts and then I thought, I didn't really need those posts in the first place. It would have been a lot easier if, instead of fooling around with those heavy pieces of timber, I had just built a stud wall to begin with, properly braced. Now I had to put the studs in, and bracing was more of a problem . . . . So, what do you think of a post and beam house as compared with stud construction?

I think that probably aesthetically post and beam is very pleasing; however, I don't think that it has any greater actual utility than the normal stud-built construction. As you just pointed out, you still have to fill in the spaces, and the post and beam construction would serve industrial applications much better than it would a house application. In industrial applications you want large spaces or spans between posts, beams, that you don't normally require in a residential construction.

There's a new book that just came out that looks like it might even be influential with people building their own homes and it shows all post and beam construction. Maybe because it's easy to visualize, people like it. Somebody that hasn't gone through the experience. A lot of people think that's the only way you should frame a house.

I personally believe they'd be making a mistake. They're wasting material where it's not needed. Like I said, it's probably aesthetically pleasing, but I don't think it is easy to build and in some respects

may be more difficult than normal, conventional stud construction.

Well, it's funny to work my way back to that after trying all these other things, to find out that the conventional way of doing it seems to be the easiest and most economical, and in a lot of ways, the most durable. It takes a lot less time to do it that way. The stud, platform, balloon, whatever method of construction was really a breakthrough, where you can build a house of very light, individual components and not have to wrestle around with large timbers, large masses.

One of the other things was that the post and beam construction frequently used masonry in-fill walls which we don't use in this part of the country because of earthquake requirements. This goes way back to the old "half timber" designs that you find in Europe, probably one of the best examples of that. For residential construction, the old, conventional stud framing is probably the most economical and cost-effective type of construction . . . .

From what you've seen, do you think that anyone can build a house?

Yes, I think that almost anyone can build houses. We've issued permits to women to build their own houses and certainly if a very slightly-built woman can build a house there shouldn't be a problem with anyone else.

I wasn't thinking so much in terms of strength as just in terms of getting everything together, in doing it in such a way that they're not wasting materials.

That's not something that's acquired readily, because when people design something they're not generally aware of the trade sizes of lumber, and they may, in fact, design a building that's quite inefficient in terms of saving materials. In fact, architects who are more concerned with aesthetics frequently ignore some of the savings that might be made

by making a room either six inches smaller or six inches larger, or a foot smaller, a foot larger, depending on the exact configuration . . . .

If someone's building a house without much experience and without going through the code, what do you think would be the most likely mistakes they'd be making?

Probably, so to speak, the thing that holds any building together, is the connections that are made. People generally aren't stupid; they use some common sense but they're not familiar with how various materials need to be connected together. And, of course, they're generally not quite knowledgeable in terms of how much the particular piece of wood will carry in terms of roof rafters or floor joists or beams.

Connections and loads.

But the connections are vital in holding the building together.

Connections such as anchor bolts, you mean?

Anchor bolts, nailing.

Type of nails and spacing?

Type of nail, spacing, length of the nails, seeing that when they are installed they don't split the lumber, that the connections not all be made at one point. For example, we find that people who do not know will cut all their top plates at one point and of course there's no lateral connection between the plates . . . . In the code it calls for lapping these four feet . . . . We recognize that sometimes there's no way that you can lap them four feet, for example, at the intersection of closets, where the closet may only be thirty inches wide. So, here we need to use some common sense. But it's still possible to make a good connection.

And bracing?

Bracing is again, a vital area.

Do you think the codes have discouraged owner-builders lately?

No, I can't go along with the idea that the codes are too restrictive. There may be some particular points of the code that may be too restrictive. But generally, I think they have provided us with a real good minimum level. That's what they are. They're not a guide. Many people think of them as a guide. They're not. Example: handrails. For many, many years the code provided that handrails must not be less than 36" high. But statistics have proven that people fall over 36" high handrails. Especially in industry. So they raised the height of

continued

handrails to 42" because that is above the human body. Seldom do you fall over one accidentally.

What is the building inspector faced with as far as giving advice or enforcing the code?

The building inspector serves only a ministerial function in that he has no discretion or very little discretion. The laws or the codes are adopted by the legislative body and then it's his job to enforce those regulations. He has the responsibility to see that the structures are safe from the structural standpoint, from the health standpoint, fire standpoint. So he has an obligation to enforce the regulations. The liability, if you will, responsibility, authority, are all spelled out fairly plainly in the . . . Code. It tells what his duties are, what his responsibilities are. It's very difficult to put into words the feeling of the building official who knows that he has sometimes as many as several hundred thousand people who may be affected by any action or inaction that he may take. Inaction is as dangerous in some respects as action . . . .

Do you think it varies as to how far you get away from an urban area, you know, the more you're away from the city . . . .

Yes, I think that probably as you get away from the urban areas you could relax, not the structural part of the code . . . . But you might relax some of the other requirements such as are presently in the State Housing Act which says if you have a residential premise within 300 feet of a serving electric utility you shall connect . . . you shall have electric wiring, and so forth. I would require the electric wiring in an urban area. I would be much more prone to allow it to be constructed without electricity in a rural area. In an urban area, if we have these 270,000 people in Marin County all burning kerosene lamps, or candles, I can see a tremendous fire problem. But if it's out on a ranch somewhere and surrounded by plenty of open space so that a fire might affect the individual family but would not affect the general welfare to an extent to where other families would be endangered, I could eliminate the need for electrical wiring . . . .

You talk about protecting people, that's what we're talking about. This is building inspection.

Building inspection is to protect the public, right?

To protect the individuals and the public, in that they are interrelated. You cannot separate one from the other . . . if a man has a large tract of land and he lives out in the center of this piece of farm land there is less chance that he will transmit disease to his neighbors. However, there's one thing that knocks the hell out of this and that is that he now has mobility. Years ago, when I was a kid for example, it was common for farmers to have an outhouse. However, they also had horses and buggies. So that infection, if it developed, traveled slowly. I mean 30 miles a day was a good day. Today it's not uncommon for a person to drive a thousand miles in one day. And to be a carrier of disease it would not be uncommon to carry that disease for a thousand miles in 24 hours. So the same parameters that were true 50 years ago are no longer true. Or 40 years ago . . . .

There are many books on the market which deal with home construction. The only thing they don't say in the books, generally, is if you're in doubt consult your local building official.

Yeah. And the building official should cooperate.

Oh. That's why we're here. We're servants really. We're providing a service. So we're a service agency. We have no private axe to grind. I treat one person, I treat the home builder who's building a deck on his house the same way I treat a guy who's building a high rise. I expect the fellow who's building the high rise to be more sophisticated and have a greater knowledge. But I still treat them the same . . . .

(Later, while talking about using local materials:)

. . . you're not going to find too many pieces of property that have enough wood on them to build a house with. It's fine to talk about using your own stuff . . . .

I didn't mean that so much as just the fact that nearby you'd find things. As opposed to buying things that are trucked in great distances. It seems that things are getting outrageously expensive and difficult to get . . . .

If people want to build their own houses, and they want to build them the minimum size, I think that you'll agree that 300 square feet is not a very big house. And that provides for one bedroom. Now — 370 square feet providing for 2 bedrooms — that's a very small house. I believe that most people could afford a house like that.

Yeah, I see. You mean even buying plywood or whatever.

Buying whatever.

That's a good point.

The thing is that . . . how many people choose not to live in a house that small?

You could build a real simple little house, couldn't you?

Definitely.

With possibility for expansion. You could build a 300 square foot house with a kitchen and bathroom in it and just build it out of studs and no big spans and have it be simple with the possibility to expand.

Any way you want. □

Construction

Introduction

In the preceding pages we have outlined the planning and design of small wood frame buildings. The following section, pages 112 - 134 is an *introduction* to stud construction of small buildings.

This is not meant to be a building manual, nor do we recommend you use these instructions alone to attempt any of the phases of house construction. If you intend to build, we suggest you read through these pages; then that you consult:

1. a good carpentry book such as *Fundamentals of Carpentry, Vol. 2 — Practical Construction* by Walter E. Durbahn and Elmer W. Sundberg (see bibliography). This is the book used by most carpentry union apprentice programs.

2. the building inspector and local building codes as to local requirements and special conditions. Even if a permit is not required, the inspector and codes can be useful in alerting you to conditions in your particular area, such as frost, snow loads, hurricanes, earthquakes, strong winds, high water table, etc.

3. a local builder if possible, for experienced advice.

We do not mean to imply that any one method of construction (rectangular wood-stud frame in this case) is appropriate in every situation. But we do believe that stud framing is the most practical, economical, durable method of constructing a building by hand in *most* building situations in North America today. And even when other materials such as adobe, stone or brick are used for walls, the roof will generally be of wood frame construction.

The Origin of Stud Construction

For hundreds — perhaps thousands — of years wood buildings were framed with heavy timbers, often a foot square, that were mortised and tenoned, then fastened together with pegs, then raised into position with group labor. In the mid-1800's the power sawmill became more efficient and began turning out lightweight 2x4's, 2x6's and other structural members, and a radical new method of construction was born: the nailed-together stud frame. This meant a lighter, stronger, quicker method of construction that is still the most common building system in North America today.

Foundations

First then concerning the Foundation, *which requireth the exactest care; For if that happen to dance, it will marre all the mirth in the House . . .*
Sir Henry Wotton,
The Elements of Architecture, 1624

Choice of foundation type depends upon site, soil, climate, building codes, available material and builder's skills.

Concrete Perimeter (Wood Floor)

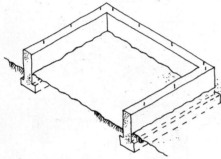

Advantages:
— conforms to most local codes
— no engineering required
— monolithic, good in earthquake
— crawl space protected from weather and debris
— good in most soils
— good on level sites

Disadvantages:
— careful formwork essential
— considerable excavation and site disturbance to get below frost line
— requires "readimix" truck or on-site concrete mixing
— the steeper the site, the more difficult

Concrete Piers

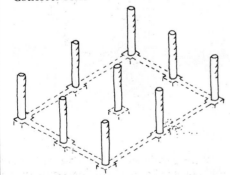

Advantages:
— miminal excavation and formwork
— good on steep sites

Disadvantages:
— leaves underside of house open
— engineering often required for code

Pre-Cast Piers

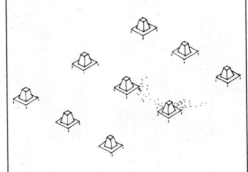

Advantages:
— quick, cheap, easy
— lightweight
— o.k. for small buildings

Disadvantages:
— open under floor
— poor anchor to ground
— no strong interconnected footings

Concrete Slab

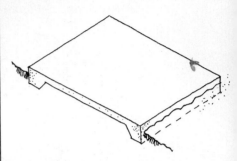

Advantages:
— all of above, plus:
— floor close to ground level
— fireproof
— no floor or foundation rot
— can be used for passive heat storage or radiant heating (pipes inside slab)
— water will not damage floor; easy to wash

Disadvantages:
— no good on steep site
— damp and cold if improperly built
— requires fill where high water table exists
— requires considerable manpower and skill to pour and finish

Wood Poles

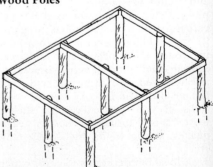

Advantages:
— good on slopes, steep sites, or where no concrete available
— minimal site disturbance
— no concrete forms
— good in remote areas

Disadvantages:
— poles eventually rot (said to last 75 years)
— not good on soft, wet soils
— deep holes may be difficult to dig
— leaves underside of house open

Laying Out a Foundation

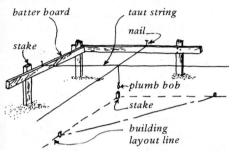

batter board *taut string*
nail
stake
plumb bob
stake
building layout line

- Shown in the diagram are batter boards: they are set back about 4' from the building lines to support strings; the strings can be removed when working (digging, building forms) then reset to check measurements.
- Accuracy in layout is essential to avoid later problems.

Procedure:
- Using 3-4-5 (6-8-10) triangle to get right angles, lay out rough outline of building with corner stakes.

- drive three stakes about 4' back from corner stakes for batter boards. With builder's level (rented), water level or line level, mark stakes all at same level.
- nail boards (1 x 4's) onto stakes, all level with marks.
- brace all boards.
- lay out exact measurements of building, using nylon string with loops, stretched between nails on batter boards. Use 3-4-5 triangle to get four strings in place; then check diagonals and adjust strings until measurements are accurate.
- on sloping sites, high batter boards need bracing.

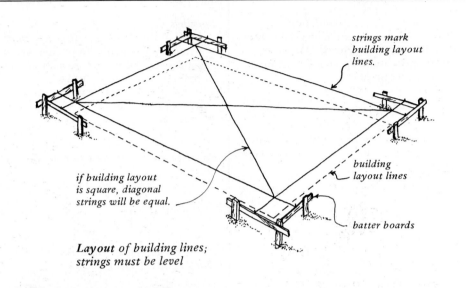

strings mark building layout lines.

if building layout is square, diagonal strings will be equal.

building layout lines

batter boards

Layout of building lines; strings must be level

Line Level must be in center of string

Builders Level

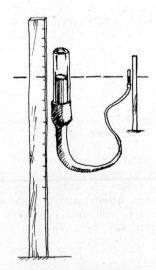

Water Level: use garden hose with short section clear vinyl tubing taped on to each end (must not leak) or you can buy the tubing with hose clamps in building materials store. Cap one end when moving. Ends must be open when reading level. Water will seek its own level. Accurate to within 1/16" in 100'.

Foundations
Building Instructions

Concrete Perimeter Foundation

This is the most common type foundation found in North America; it meets most building codes with no engineering required.

The basic parts of a perimeter foundation are the footing, which spreads the weight of the building uniformly over the ground, and the stem wall, which supports the floors and walls above ground level.

Note: interior bearing walls supporting roof load must be supported by continuous footing with stem wall and mud sill, or pre-cast piers and girder.

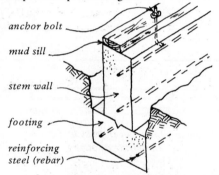

anchor bolt
mud sill
stem wall
footing
reinforcing steel (rebar)

No. of stories	Thkns. of Fnd. wall in.	Width of ftng. in.	Thkns. of ftng. in.	Depth below fin. grade (in.)*
1	6	12	6	12
2	8	15	7	18
3	10	18	8	24

Bottom of footing must always be below frost line.

Inexperienced builders may find it easier to first pour the footings, then build forms for stem walls on top of footings after concrete has set. Experienced builders generally pour footings and stem walls in the same pour.

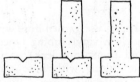

Left: *footing with key*; middle: *stem wall poured on top of footing*; right: *footing and stem wall poured same time.*

Footings

— cold climate concrete footings must extend 12" below frost line or to solid rock. Otherwise freezing water can expand and break concrete wall. If you are building in cold climate, ask natives how deep frost penetrates.

Footings must bear on undisturbed firm soil

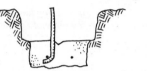

Firm soil: no form boards required. Pour concrete directly in trench.

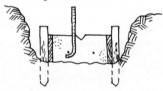

Loose soil: form boards required. A 2 x 2 is pressed into the wet concrete.

Hillside site: footing and stem wall are stepped:

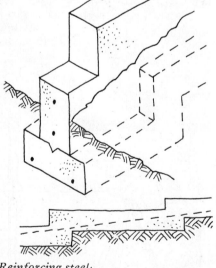

Reinforcing steel: shallow steps, as above; steeper steps, as below

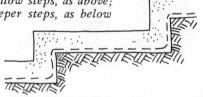

Excavating: using strings, mark lines for digging on ground with lime or white flour. Careful digging can save concrete. Throw dirt well back from building lines so you'll have room to work. Get bottom of trenches level with carpenter's level on 2x4.

Pouring: Steel should be in place: 3" up off ground, at least 2" from sides of wood forms. In some areas you can buy *dobies*, a 3" square of concrete with wires to hold steel off ground. Wet earth just before pouring so dry ground won't suck moisture out of concrete.

Dobie

After pouring, smooth off and level top of footing (upon which form walls will rest). Put 2 x 2 on edge in wet concrete for key. Be sure there is vertical steel (hooks under horizontal steel) every 18".

Building Forms

The forms for the stem wall are built on top of the footing, which has a key and vertical steel to tie footing and wall together. Forms are filled with concrete to top; thus forms must be accurate in height as well as horizontal measurements.

The 2" floor joists are commonly used for form lumber, later stripped and used as joists.

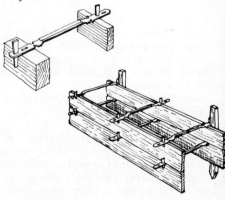

Steel form ties are used to hold forms together while under concrete pressure. They remain in concrete; ends are broken off after forms are stripped.

Building forms on top of already-poured footing

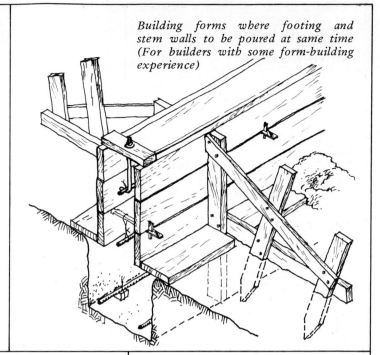

Building forms where footing and stem walls to be poured at same time (For builders with some form-building experience)

Procedure:
- using strings, build outside forms first. Use builder's level or water level to make sure they are level all around. Use blocks or stones to raise forms if necessary.
- check diagonals, adjust if necessary. Accuracy is important here!
- insert steel ties.
- bend horizontal steel and lay in before adding inside form.
- build inside forms.
- raise and tie steel as forms are built.
- suspend anchor bolts in place.

Reinforcing Steel
Two or more (depending on wall height) ½" bars of reinforcing steel ("rebars") are hung inside forms with tiewire from the steel ties. See drawing, above right.

Anchor Bolts
Must be in place before pour. See p. 114.

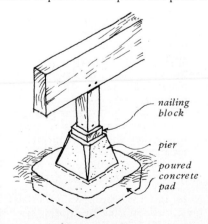

nailing block

pier

poured concrete pad

Pre-cast concrete pier set in wet concrete supports girder which carries floor joists.

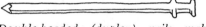

Double-headed (duplex) nails make stripping forms easier.

Pouring Concrete
Mixing it yourself: 3 parts gravel; 2 parts sand (not river or beach sand); 1 part cement. The less water, the stronger the concrete. Keep it stiff.

Readimix trucks: much easier, more expensive.

Procedure:
- things move fast and it's hard work. Have 3 - 4 helpers with as many shovels. Rubber boots, rubber gloves, wood floats and a big wheelbarrow.
- spray ground and forms with water just before concrete is poured.
- place in layers; pour bottom of form all the way around first. This gives it time to set up, minimizes pressure and leaks.
- if concrete starts oozing from form, it can often be stopped by shovelling dirt against the side of the form that's leaking, then pouring elsewhere for a while.
- tamp or puddle concrete in forms with a 1 x 4 to ensure good fill, no voids.
- tap side of forms with hammer for smooth surface.
- when form is full, trowel off the top smoothly. (Mud sill will rest here.) Check anchor bolts for right height, vertical angle.
- hose off all tools immediately.
- strip forms after 24 hours. Keep concrete covered and damp with plastic or burlap for 3 - 5 days.

Forms are suspended over trench as above, or stakes are driven in to trench for vertical support, then removed as soon as concrete has set (some building inspectors will not allow stakes in trenches).

Openings in concrete:
Must be blocked out in formwork before pour.

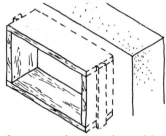

Typical rectangular opening with redwood stringers left in.
Crawl space: so you can get underneath house.
Vent openings: Good ventilation minimizes termite and dry rot damage. Building codes require 1½ sq. ft. for each 25 lineal ft. of exterior wall.
Plumbing and wiring openings: check with a plumber for sewer pipe (must be inserted at proper angle) and electrician if underground service is planned.
Girder supports:

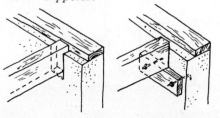

Left: *pocket formed by blocking out inside form;* right: *anchor bolts hold treated wood support.*

Foundations

Building Instructions
continued

Basements

The advantage of a basement is the extra space under the same roof. Disadvantages: for high walls, you should get help from an experienced builder and/or engineer; also, basements are not advisable in wet, flat areas (no drainage away from bldg.).

If you are considering a basement, look at other houses in your area. If there are basements, they are probably practical.

Soil stability, drainage and winter freezing conditions are all important factors here. If you are considering a basement we recommend you consult a local builder.

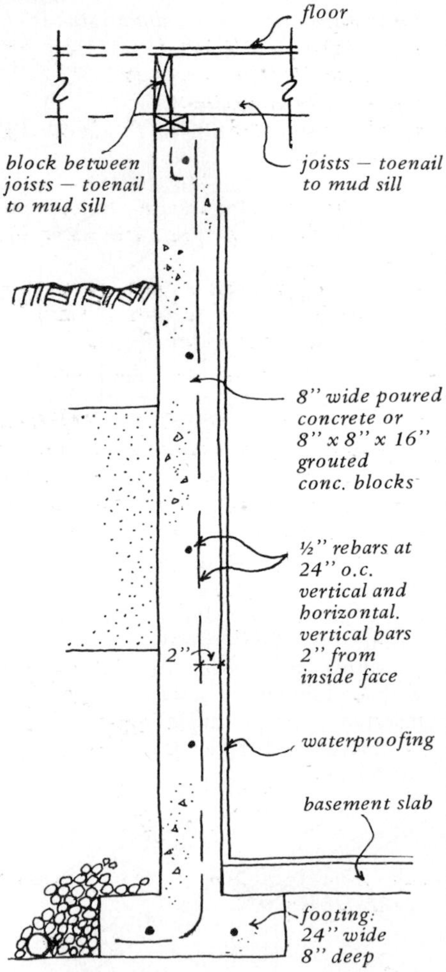

floor
block between joists — toenail to mud sill
joists — toenail to mud sill
8" wide poured concrete or 8" x 8" x 16" grouted conc. blocks
½" rebars at 24" o.c. vertical and horizontal. vertical bars 2" from inside face
2"
waterproofing
basement slab
footing: 24" wide 8" deep

Typical basement wall. Top of wall must be braced by floor joists toenailed to mud sill. Not a retaining wall.

Concrete Block

Concrete block walls can be built without forms. For strength they are usually reinforced with steel and filled with concrete. They are not as strong as concrete walls.

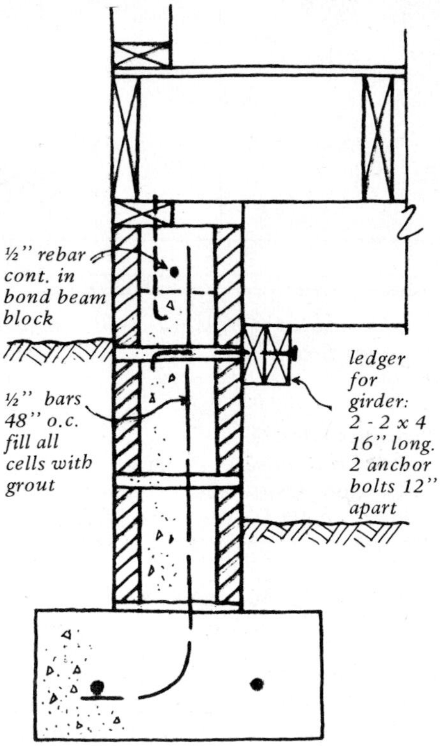

½" rebar cont. in bond beam block
½" bars 48" o.c. fill all cells with grout
ledger for girder: 2 - 2 x 4 16" long. 2 anchor bolts 12" apart

Typical concrete block stem wall

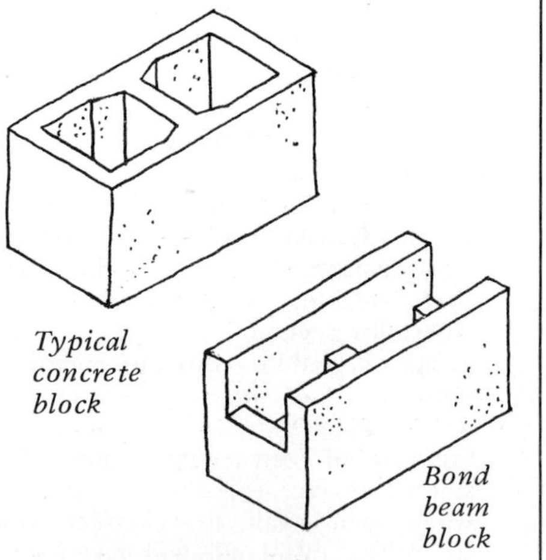

Typical concrete block
Bond beam block

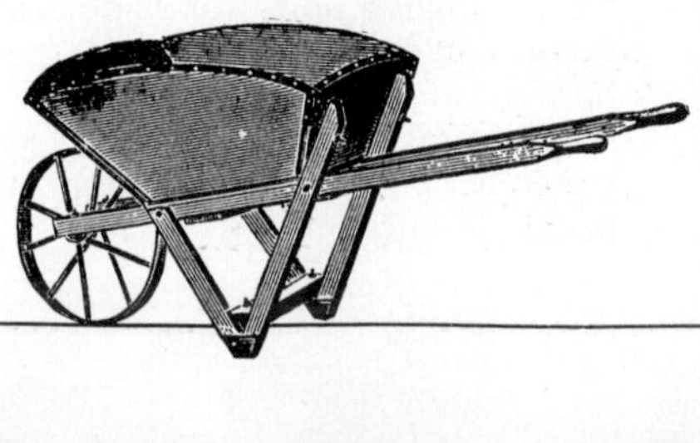

Concrete Slab Floors

If built well, a concrete slab need not be damp and cold. Concrete has a high heat storage capacity and can stabilize room temperature.

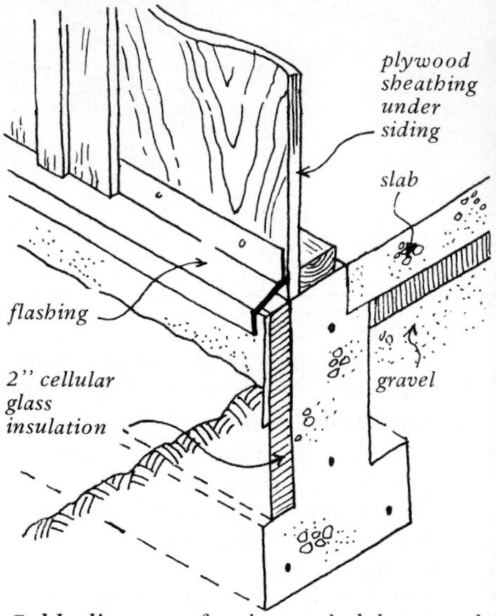

plywood sheathing under siding
slab
flashing
2" cellular glass insulation
gravel

Cold climate: footing and slab poured separately; insulation under slab. (Temperatures below 25^{0} F.)

Procedure:

- first build standard footing and stem wall up to desired height. Omit all vent and crawl hole openings but allow for sewer, water and electrical lines, which must be installed in the floor and/or foundation wall before the pour.
- make provision, if necessary, for heating ducts or radiant heating.
- remove all organic material from floor area.
- put down 4" - 6" of 1½" gravel.
- lay down 1½" or 2" rigid waterproof insulation.
- apply waterproof vapor barrier.
- roll out 6" x 6" welded wire mesh or place 3/8" rebar, 18" on center, each way.
- install any plumbing or wiring pipes, wrapped with fiberglas insulation where they will be in concrete.
- set anchor bolts.
- set wood screeds, either temporarily or permanently.

Note: interior bearing walls on slab floor must be supported by footing same size as exterior footings.

concrete slab continued

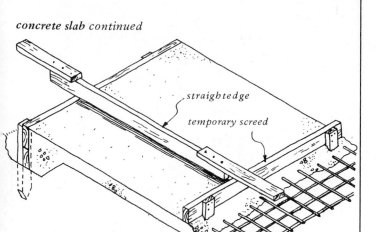

Temporary screed *is staked above slab level for easy removal after pour. Maximum effective length for straightedge is 10-12 feet; use temporary screed as above when pour is wider.*

Straightedge *is worked back and forth to bring concrete to rough finish level. After this, use wooden float, then steel trowel to finish. Pull mesh up during pour.*

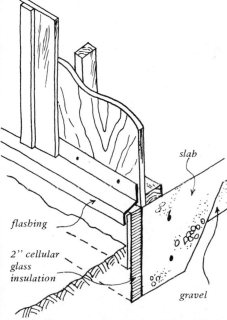

slab

flashing

2" cellular glass insulation

gravel

Mild climate: *footing and slab poured together; gravel under slab.*
(No temperatures below 25⁰ F.)

Procedure:

— using batter board strings, lay out building lines and form around outside edges with 2 x 10's or 2- 2 x 8's.

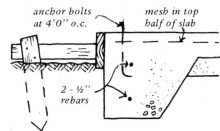

anchor bolts at 4'0" o.c.
mesh in top half of slab
2 - ½" rebars

— dig footing 12" deep (one story), 12" wide. Remove sod and organic material from floor area.
— lay down 4" - 6" of 1½" gravel.
— next, vapor barrier (not under footings).
— install anchor bolts.
— set screeds.
— put rebar in footings, wire mesh or rebar in floor area and install any plumbing or electrical pipes. Follow instructions for pouring slab as at above right.
— after stripping forms, insulate exterior of slab and backfill with gravel; install drain tiles where needed.

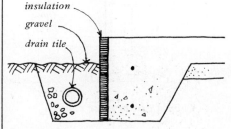

insulation
gravel
drain tile

The Pour

See *Shelter*, p. 47 for extensive instructions. Brief instructions:
— pour 3½" finish floor.
— rake off level with 2 x 4 resting on screeds.
— smooth off with large wooden float. A film of water will rise to the top. When this evaporates, it's ready for steel trowel.
— continue trowelling until surface is smooth and hard, and steel "rings".
— in hot weather, shade if possible.

Screed to Stay in Floor. Redwood screeds or rot-resistant wood can be used as guides to level concrete. They can be left in slab permanently as expansion joints or used later to lay wooden floor on top.

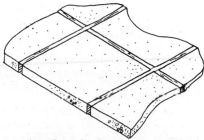

Pouring floor in 2 or more stages; make construction joint.

117

Foundations

Building Instructions
continued

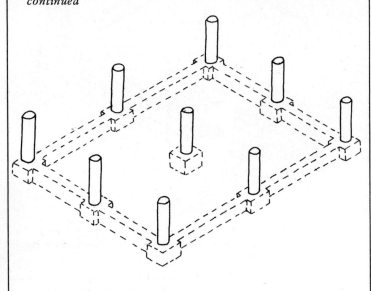

Cast-in-place concrete pier foundation

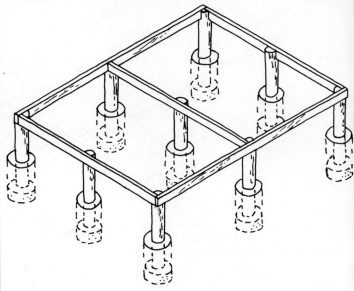

Wood pole foundation

Cast-in-place Concrete Piers

Cylindrical cardboard forms — often called *sonatubes* — are available in several diameters. Main advantages are minimal formwork, and suitability to sloping sites. Three ways to utilize:

1. With regular footing, if height of the concrete pier is less than 3 times diameter of pier, and if piers are spaced 6' apart or closer.

2. With a *grade beam,* a steel-reinforced concrete beam, at ground level that acts as a bridge to tie all piers together.

3. With deep holes and no connecting footing or grade beam; here the weight of the building is partially supported by friction of the soil around the piers. Holes are usually 6' - 12' deep, depending upon soil conditions and height of pier above ground. Engineering recommended here.

Simplified system of using concrete pier foundation:

— check with local builders and/or building codes as to footing and pier size and depth.
— dig footing trenches.
— enlarge footing trench around each concrete pier. If site is flat, forms probably not needed. If site is steep, stepping may be required (see page 114).
— lay 2 pieces steel in trench supported 3" above ground.
— wire 4 pieces vertical steel in each pier area.
— pour trenches.

— next day place cardboard tubes over pier locations. Plumb them, and cut off at desired height, brace with 1 x 4's if necessary.
— insert vertical steel to 2" of top, insert anchor straps for posts or girders.
— throw some dirt around the base.
— fill with concrete. Be sure to puddle with stick to eliminate voids.

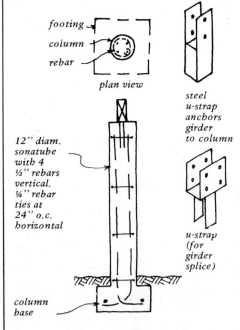

footing
column
rebar

plan view

12" diam. sonatube with 4 ½" rebars vertical. ¼" rebar ties at 24" o.c. horizontal

column base

steel u-strap anchors girder to column

u-strap (for girder splice)

Typical cast-in-place column

Wood Poles

Especially suited to inaccessible or sloping sites where concrete work is difficult. Treated poles are said to last 75 years.

Number, size of poles, depth and spacing determined by soil conditions. Depth of pole also depends on height of pole above ground: the higher, the deeper. Many local building codes require that wood pole foundations be engineered.

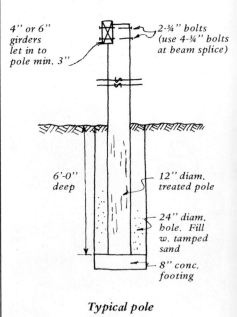

4" or 6" girders let in to pole min. 3"

2-¾" bolts (use 4-¾" bolts at beam splice)

6'-0" deep

12" diam. treated pole

24" diam. hole. Fill w. tamped sand

8" conc. footing

Typical pole

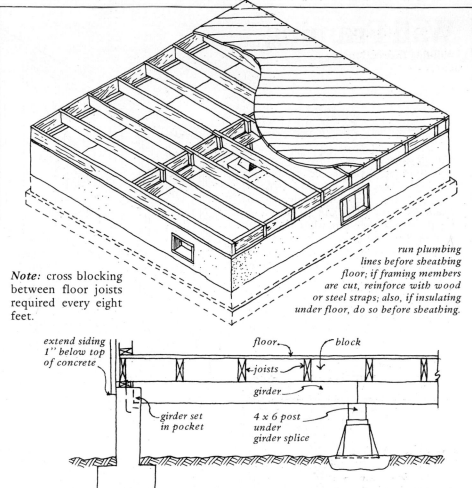

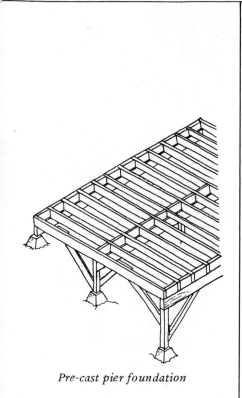

Pre-cast pier foundation

Note: cross blocking between floor joists required every eight feet.

extend siding 1" below top of concrete

floor — *block*

joists

girder

girder set in pocket

4 x 6 post under girder splice

run plumbing lines before sheathing floor; if framing members are cut, reinforce with wood or steel straps; also, if insulating under floor, do so before sheathing.

Typical floor framing

Pre-cast Piers

Pre-cast concrete piers are a quick, easy, cheap foundation for small lightweight buildings; good for barns, sheds, coops, etc. For added bearing, excavate a 24" x 24" hole, 12" deep beneath each pier and fill with concrete. When concrete is partially hard, set pier on top.

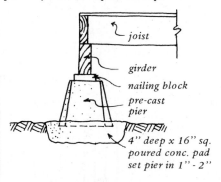

joist

girder

nailing block

pre-cast pier

4" deep x 16" sq. poured conc. pad set pier in 1" - 2"

Typical pre-cast pier

Floor Framing

Use redwood or treated wood sills; other woods will rot. Measure diagonals to make sure sills are set accurately.

Setting the sill: bolt sill down using mortar or wood shims (shingles) to level exactly, using builder's level or water level. Use mortar to fill cracks after shimming.

Girders: set girders, shim where necessary to bring top edge of girder flush with sill.

Floor joists: precise layout of floor joists is important. Mark position of all joists on sills and girders.

Measuring tip: using long tape measure, mark all joists at once; do not measure from joist to joist, as this will accumulate error.

Joists spaced 24" can be sheathed diagonally with ¾" boards, or ¾" tongue & groove plywood. Solid blocking over girder. Toenail joist to sills and girders with 3-8d or 2-16d nails. Double floor joists required under bearing partitions when running parallel with joists.

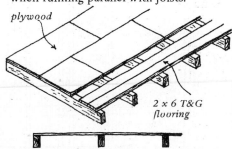

plywood

2 x 6 T&G flooring

Above:

Joists for plywood floor: measure from outside of first joist to center of next joist. Plywood must meet in center of joists, requires accuracy.

Plank and beam flooring: 4" rather than 2" material used for floor frame; covered with 2 x 6 tongue & groove, or 1-1/8" plywood. 4 x 6 or 4 x 8 joists spaced 48"

SIZE OF FLOOR JOISTS (Inches)	SPACING OF FLOOR JOISTS (Inches)	MAXIMUM ALLOWABLE SPAN (Feet and Inches)			
		GROUP I	GROUP II	GROUP III	GROUP IV
2 x 6	12	11-6	10-0	8-0	6-0
	16	10-0	8-6	7-0	5-0
	24	8-0	7-0	6-0	4-0
2 x 8	12	15-0	13-6	11-6	8-6
	16	13-6	11-6	10-0	7-6
	24	11-0	9-6	8-0	6-6
2 x 10	12	19-0	17-6	14-6	11-6
	16	16-6	15-6	13-0	10-0
	24	14-0	13-0	10-6	8-6
2 x 12	12	23-0	21-6	19-0	14-6
	16	20-0	19-6	16-6	13-0
	24	16-6	16-6	13-6	10-6

Table title: **ALLOWABLE SPANS FOR FLOOR JOISTS USING NONSTRESS-GRADED LUMBER**

Group I: const. grade, Doug. Fir, West. Larch, South. Pine.
Group II: const. grade, Sitka Spruce, White Fir, W. Coast Hemlock; spec. grade South. Pine; No. 1 grade East. Spruce.
Group III: Standard grade Fir, Hemlock, Larch, Spruce.
Group IV: Utility grades any of above.

Wall Framing
Building Instructions

The advantage of platform, or western type framing (as opposed to balloon framing, where studs extend two full stories) is that studs and plates can be assembled into walls and partitions on the floor and tilted into place. The wall is framed on the floor with a *single* top plate; after the wall is erected, the 2nd top plate is added to tie walls together. Note how plates lap at corners to tie separate wall units together.

Actual Lumber Sizes

1 x 4 = ¾" x 3½"		2 x 8 = 1½" x 7¼"
2 x 4 = 1½" x 3½"		4 x 10 = 3½" x 9¼"
2 x 6 = 1½" x 5½"		

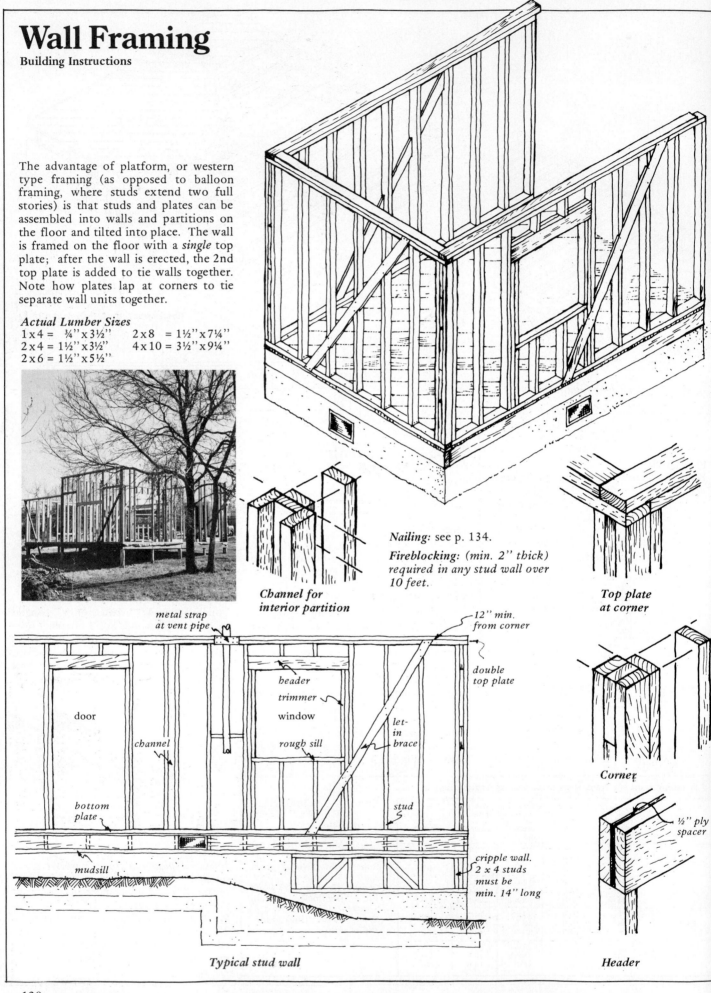

Channel for interior partition

Nailing: see p. 134.

Fireblocking: (min. 2" thick) required in any stud wall over 10 feet.

Top plate at corner

metal strap at vent pipe

12" min. from corner

header

trimmer

window

rough sill

door

channel

let-in brace

double top plate

bottom plate

stud

mudsill

cripple wall. 2 x 4 studs must be min. 14" long

Corner

½" ply spacer

Typical stud wall

Header

120

Tilt-up Wall Layout Procedure:

- with chalk line snap line 3½" back from floor edge.
- lay top and bottom plates along line. Nail together (temporarily), 8d nails.
- lay out doors, windows, partitions. See pp. 126-127 for size of door and window openings.
- at joints in plates, mark for 2 studs; stagger joints on top and bottom plates — not one over the other.
- take top and bottom plate apart.
- tack bottom plate to floor flat (temporarily), and lay studs against it.
- nail together top plate and studs.
- nail bottom plate on.

With solid or let-in bracing, wall can be braced before tilt-up. Measure diagonals, rack into position. When square (diagonals equal), brace.
- with helpers, tilt into place.
- with string, straighten wall and brace.

Wall Bracing:

Stud walls *must* be braced. Properly nailed plywood sheathing is strongest; let-in or cut-in bracing is most common and must be placed at not more than 60° nor less than 45° from the horizontal. Building codes require that all exterior walls and main interior partitions be braced at each end, or as near thereto as possible, and at least every 25 feet of length.

tilting up wall

If wall is to be braced with sheathing, a temporary brace will hold wall in place until tilted up and sheathed.

To straighten wall, nail 1" blocks each end, stretch string, hold 1" block at plate. Straighten wall, nail temporary brace to secure wall.

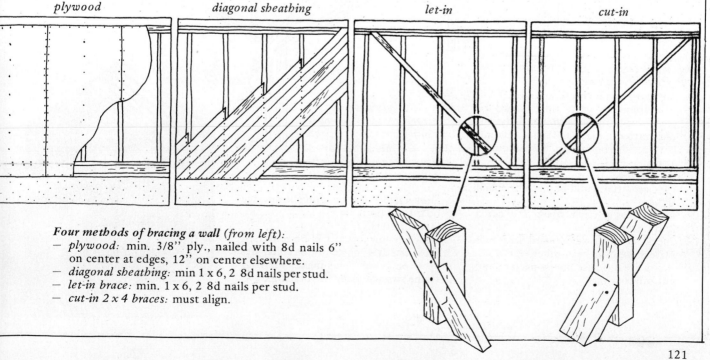

plywood   diagonal sheathing   let-in   cut-in

Four methods of bracing a wall (from left):
- *plywood:* min. 3/8" ply., nailed with 8d nails 6" on center at edges, 12" on center elsewhere.
- *diagonal sheathing:* min 1 x 6, 2 8d nails per stud.
- *let-in brace:* min. 1 x 6, 2 8d nails per stud.
- *cut-in 2 x 4 braces:* must align.

Roof Framing

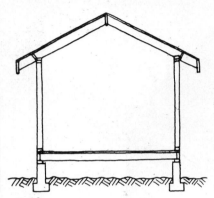

Gable
Shorter spans than flat or shed; can have loft or 2nd story.

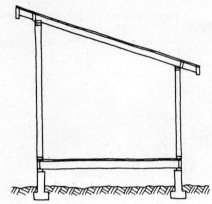

Shed
Simple, can have ½-loft one side.

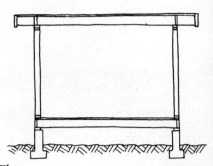

Flat
Simple, low profile, can add 2nd story later. Not good in snow country.

Shown above are the three simplest roof shapes. Choice of a shape depends upon weather conditions (steep slope for heavy snow country), overall size of building, materials available, and builder's skills. See pp. 88-101 for variety of roof shapes.

Gable Roof Building Instructions

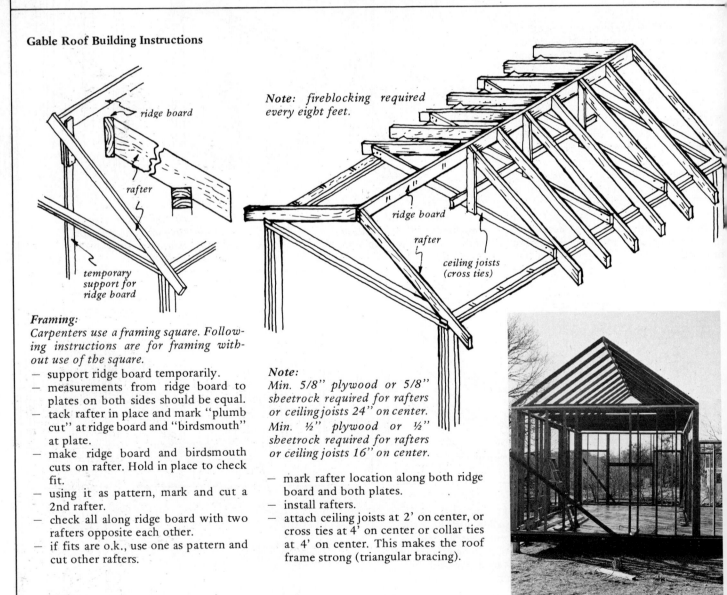

ridge board

rafter

temporary support for ridge board

Note: *fireblocking required every eight feet.*

ridge board

rafter

ceiling joists (cross ties)

Framing:
Carpenters use a framing square. Following instructions are for framing without use of the square.

— support ridge board temporarily.
— measurements from ridge board to plates on both sides should be equal.
— tack rafter in place and mark "plumb cut" at ridge board and "birdsmouth" at plate.
— make ridge board and birdsmouth cuts on rafter. Hold in place to check fit.
— using it as pattern, mark and cut a 2nd rafter.
— check all along ridge board with two rafters opposite each other.
— if fits are o.k., use one as pattern and cut other rafters.

Note:
Min. 5/8" plywood or 5/8" sheetrock required for rafters or ceiling joists 24" on center.
Min. ½" plywood or ½" sheetrock required for rafters or ceiling joists 16" on center.

— mark rafter location along both ridge board and both plates.
— install rafters.
— attach ceiling joists at 2' on center, or cross ties at 4' on center or collar ties at 4' on center. This makes the roof frame strong (triangular bracing).

Sheathing

Rafters at 24" spacing can be sheathed with ¾" lumber or ½" plywood. Sheathing over open cornices should be boards — can be plywood over roof. This way, no ply exposed to weather or visible from underneath.

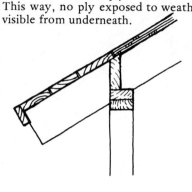

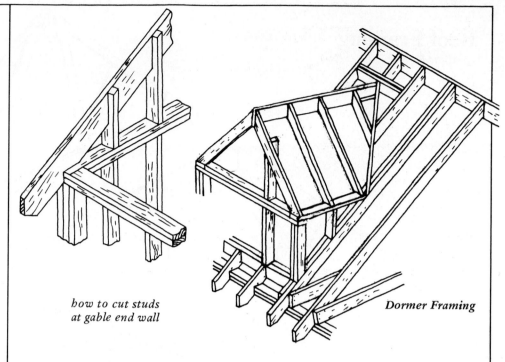

how to cut studs at gable end wall

Dormer Framing

Eaves

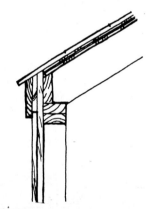

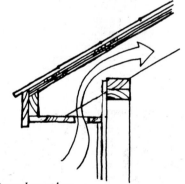

Open cornice:
Rafter overhang exposed, blocking vented and screened. For gutter cut rafter end plumb.

Closed cornice:
No rafter overhang; cheap and fast. Good for heavy winds, but no rain protection for walls, no overhang protection for vent holes. Gutter can be nailed to fascia.

Boxed cornice:
Rafter tails boxed with screened vent in soffit. Gutter nails to fascia. (Vents should not be over, or within 3' of windows or doors.)

Flat Roof

Good for adding 2nd story later: use floor joist-sized rafters — cheap 2 ply tar and gravel roofing. Frame in stair opening and cover with skylight; later move skylight above stairs when adding 2nd story.

If no 2nd story planned, it's best if roof has slight pitch. Add an extra plate to one side of wall — gives 1½" slope.

Disadvantage: heavy framing required for snow loads. (Snow slides off steeper roofs).

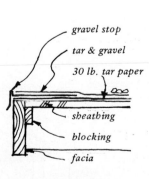

gravel stop
tar & gravel
30 lb. tar paper
sheathing
blocking
facia

Flat Roof Eave

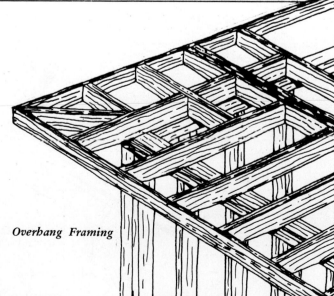

Overhang Framing

Roof Framing
continued

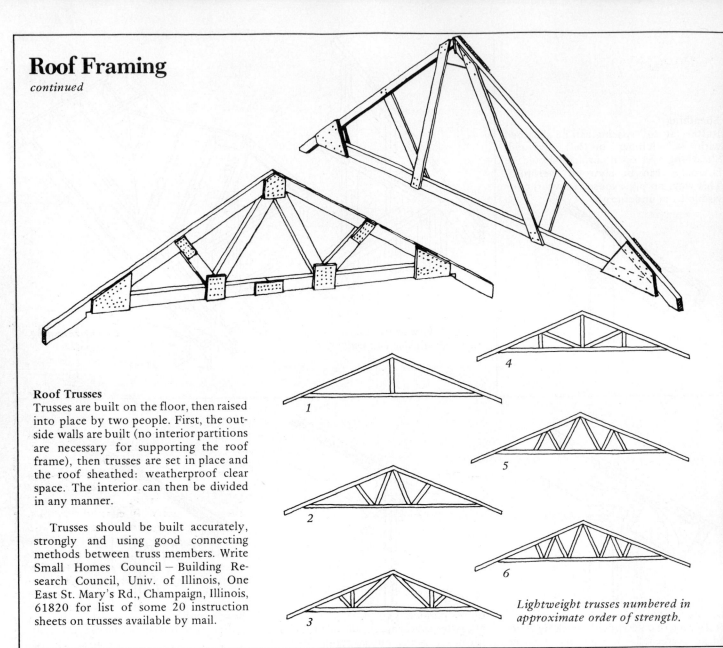

Roof Trusses

Trusses are built on the floor, then raised into place by two people. First, the outside walls are built (no interior partitions are necessary for supporting the roof frame), then trusses are set in place and the roof sheathed: weatherproof clear space. The interior can then be divided in any manner.

Trusses should be built accurately, strongly and using good connecting methods between truss members. Write Small Homes Council – Building Research Council, Univ. of Illinois, One East St. Mary's Rd., Champaign, Illinois, 61820 for list of some 20 instruction sheets on trusses available by mail.

Lightweight trusses numbered in approximate order of strength.

Gutters

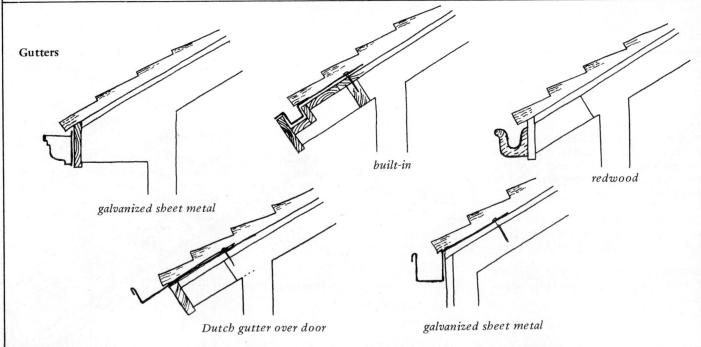

galvanized sheet metal

built-in

redwood

Dutch gutter over door

galvanized sheet metal

Roofing

Common Roofing Materials:
- 90 lb. mineral paper roll roofing
- 240 asphalt shingles
- tar and gravel
- wood shingles
- wood shakes
- concrete or ceramic tiles
- aluminum roofing
- fiberglass
- Corten steel
- corrugated steel

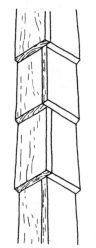

*finishing ridge —
note alternate
overlaps*

Wood Shakes

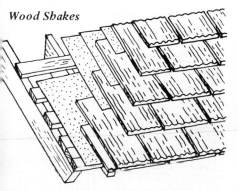

Similar to wood shingles, but longer, thicker, split rather than sawn (often sawn on one side). Shakes don't lie as flat as shingles due to rough texture. 30 lb. felt liner is installed between courses. See *Shelter*, p. 61, for shake instructions.

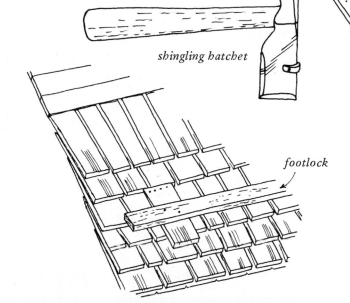

shingling hatchet

footlock

Footlock shown above is held by shingles spaced 6' apart. Provides movable foot support while shingling.

Wood Shingles
A good shingle roof can last from 20 - 70 years, depending on application, exposure and climate. Not good in hot, dry areas. Shingling a large roof is a lot of work — get helpers.

Shingling tips:
- vapor barriers not recommended under shingles; let them "breathe."
- 3"/12" is minimum recommended slope.
- nail with 3d hot dip galvanized box nails, 2 nails only per shingle, ½" either side.
- allow ¼" gap between shingles for expansion. If butted tight, roof can swell and buckle.
- use 1 x 2 board to butt shingles against and align each course.

metal edge

rolled edge

Roll Roofing
Easy, fast, cheap. Lasts 2-5 years in full sun, 8-15 years in shade or north side. Can be used as quick first roof, then covered with better roof material before it begins to disintegrate. Disadvantage: can be ripped off in high winds. Roll roofing comes with clear instructions. Install on warm day. Two people are necessary to do a good job. Roll out roofing on ground, cut to length allowing extra for trim. Let lie for a few hours in sun. Roll back up lightly, install according to instructions: stretch tightly before nailing. Seal joints with roof patch (thicker than lap cement and easier to use). Nails 2" - 3" apart.
- eaves and gable: overhang ¼", tar and nail.
- ridge: cover with 12" wide piece, tar and nail.
- in high wind areas use metal flashing to keep roofing from blowing off.

Asphalt Shingles
Low cost roofing, easy to install, last longer than roll roofing. In high wind areas self-locking shingles are used: after installation, the tar inside melts slightly, sealing them together tightly. Lay per manufacturer's instructions on bundles. Tip on cutting: score shingles on back side, bend and shingle snaps off — clean break.

Tar and Gravel
Hot tar and gravel is best left to professionals. Cold tar may be available — check with roofing suppliers.

Windows

Aluminum Windows

Made in two finishes: bright aluminum and anodized bronze (cost about 30% more).

Advantages:
— easy, fast to install
— no air or water leaks
— durable: no painting or puttying
— won't swell or stick in wet weather
— built-in screens for flies

Disadvantages:
— not good in cold climates; frost forms inside.
— interior condensation rots sills.
— aesthetics.

Good compromise: set all fixed glass (non-opening) in wood frames; use aluminum for all openers.

Installing aluminum windows: for new construction, aluminum windows have a nail-on flange which goes under the siding, i.e., nailed onto the studs.
— size of framing openings: varies with manufacturer. Some make windows 3/8" smaller than called-for size (e.g., a 36" x 48" window would actually measure 35-5/8" x 47-5/8"); others are made exactly to called-for size. Best is to have window on hand when framing, or make opening ½" wider all around than called-for size. Easy to shim opening but difficult to make it larger.

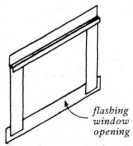

flashing window opening

— trimmers not necessary for aluminum windows.

Flashing aluminum windows:
— staple strip of paper across bottom flange first; then sides, then top, overlapping each layer (like shingles). 15 lb. tar paper goes under bottom and side strips, over top strip.
— measure diagonals to make sure window is square before nailing on. Otherwise sliders may stick. Same for sliding doors.

Wood Windows

Advantages:
— good looks.
— small panes; cheap to replace if broken.
— can often find cheap or free used.

Disadvantages:
— must be maintained: putty, paint.
— more likely to leak than aluminum.
— more time to install.

Building frames and sills:
— use clear, dry material.
— make inside of frame at least 3/16" larger than actual sash dimensions — gives clearance to install sash and room to open if window is hinged.
— use 2 x 8 clear stock for sills. To rip, use table or radial arm saw or take to cabinet shop.
— use 1 x 6 for jambs.
— cut out parts and assemble with nails and glue (drill if necessary to prevent splitting). Use waterproof (dry powder) glue.

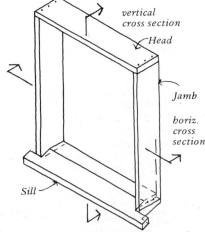

vertical cross section
Head
Jamb
horiz. cross section
Sill

— completely assemble frame before installing.
— place frame in opening, shim tight with shingles.
— check diagonals and nail through jamb into studs.
— add stops, stool, apron, trim last.

Exterior trim:
— must be done correctly to avoid leaks.

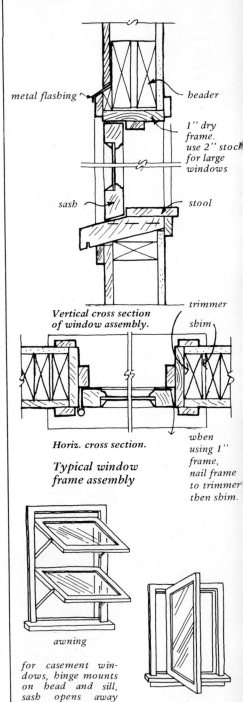

metal flashing
sash
header
1" dry frame. use 2" stock for large windows
stool

Vertical cross section of window assembly.

trimmer
shim

Horiz. cross section.

when using 1" frame, nail frame to trimmer then shim

Typical window frame assembly

awning

casement

for casement windows, hinge mounts on head and sill, sash opens away from hinge jamb.

Window with self-adjusting hinge stays where you put it. (Whitco Co., Sausalito, Calif.)

126

Doors

Door Types

- *Factory-built solid core:* plywood veneer over particle board core. Good stable doors if well sealed with paint. Medium price.
- *Factory-built panel doors:* variety of styles but most are quite expensive.
- *Used:* the best deal for owner-builders. Old doors are often well built.
- *Homemade:* weatherproof doors that won't sag or warp must be properly designed and built of sound materials.

Homemade Doors

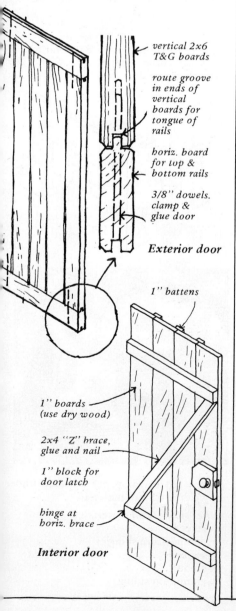

vertical 2x6 T&G boards

route groove in ends of vertical boards for tongue of rails

horiz. board for top & bottom rails

3/8" dowels. clamp & glue door

Exterior door

1" battens

1" boards (use dry wood)

2x4 "Z" brace, glue and nail

1" block for door latch

hinge at horiz. brace

Interior door

Jambs

In general, there are two kinds of jambs:

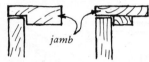

jamb

At left, rabbeted jamb which is better for heavier doors and exterior; more weatherproof, durable. At right, nailed-on stops which are easier.

Procedure for hanging doors:

- before making jambs, look closely at properly hung door; note hinge positions on door and jamb.
- use solid, dry material. Rip to correct width and rabbet on table saw or skilsaw with guide.
- lay out hinges on door first.

- use hinges as pattern, mark with sharp knife.
- mortise out with sharp chisel or router.
- lay hinge in mortise, mark screw holes; drill holes for screws.
- drill size = shaft of screw.
- Phillips head screws are easier to drive.
- screw hinges to door.
- lay jamb next to door and mark jamb for hinges (allow 1/8" clearance between top of door and head jamb). Mortise jamb. Screw hinges to jamb.

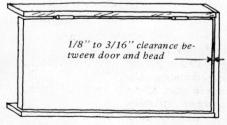

1/8" to 3/16" clearance between door and head

- assemble jambs on floor; nail and glue; let dry 24 hours.

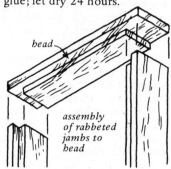

head

assembly of rabbeted jambs to head

- assemble door and jambs; check for fit and clearance.
- remove hinge pins; lay door aside.
- before finally hanging door, bevel edge on door knob side about 1/16"
- shim jambs in rough opening using shingles. Can use level, steel square, tape measure to check for plumb and square.
- nail through jamb with 8d finish nails. Nail only hinge side first.
- hang door, check for fit.
- adjust jambs for 1/16" door clearance, then nail all jambs.

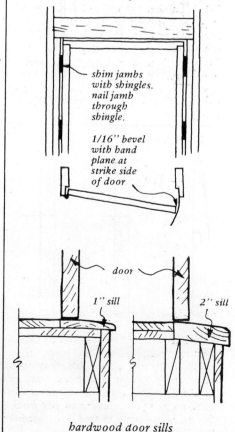

shim jambs with shingles. nail jamb through shingle.

1/16" bevel with hand plane at strike side of door

door

1" sill

2" sill

hardwood door sills

Siding

Plywood

The strongest bracing, quickest to apply. Must be exterior grade plywood, not merely plywood with exterior glue, which has interior voids.

Various woods, textures, patterns are available. To apply, make sure building is temporarily braced, square and plumb. Typical plywood siding is 3/8"-5/8" thick. 3/8" requires horizontal blocking as required for vertical board siding (see drawing in middle column). Batten joints or use shiplap plywood.

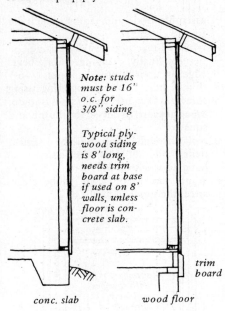

Note: studs must be 16" o.c. for 3/8" siding

Typical plywood siding is 8' long, needs trim board at base if used on 8' walls, unless floor is concrete slab.

conc. slab

wood floor

trim board

Board and Batten

Use lumber known to be stable in weather exposure: redwood, cedar, some pines, etc. Boards should be reasonably clear, no loose knots. If green lumber, stack with spacer for a few months before using.

Use two rows blocking (min. 2'-8" o.c.) between top and bottom plates for nailing. 15 lb. tar paper under siding; important: shingle tarpaper, top over bottom. Use 8d hot dip galv. box nails.

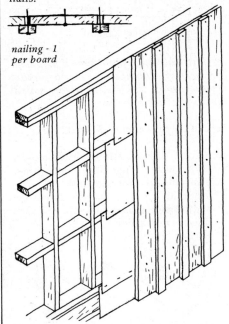

nailing - 1 per board

Shingle Siding

Requires sheathing or horizontal 1x4 stringers 4" o.c., and 15 lb. tar paper underneath. To get straight line: use 1x2 wood strip, nail at base of each course with small nail, butt shingles against.

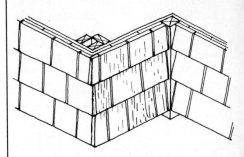

Above: *"woven" outside corner, 2 x 2 at inside corner. Below: 1 x 4 or 1 x 6 boards at outside corner.*

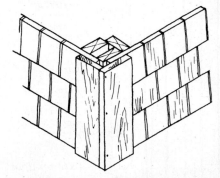

Horizontal Siding

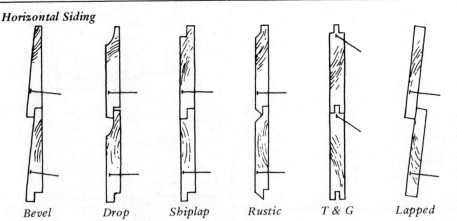

Bevel   Drop   Shiplap   Rustic   T & G   Lapped

No blocking between studs required with horizontal siding. Use hot dip galvanized nails to avoid nail stains. Use one nail per board at each stud, allows board to shrink without splitting.

Note: 15 lb. tar paper required under any siding or wall material.

Other siding/wall materials
— stucco
— composition shingles
— adobe
— stone
— aluminum siding

Insulation

Glass wool is the most commonly used insulation today. It works well in stud walls and between rafters or floor joists. Polyurethane foam, while a more effective insulation per inch, is more expensive than fiberglass (almost twice as much for equal insulating value) and highly hazardous in fires; it is used over roof sheathing where exposed rafters are desired or under or around a concrete slab floor.

Comparison of Common Insulating Materials

Material	K-factor*	R-value†
Polyurethane foam	0.15	6.7
Mineral wool	0.25	4.0
Glass wool (fiberglass)	0.27	3.7
Sound deadening board	0.37	2.7
Vermiculite	0.48	2.1
Wood-across grain	1.00	1.0

*(conductance) (Btu-in/hr. ft^2 °F)
†per inch of insulation

Definitions:

BTU – British Thermal Unit: heat required to raise one pound of water one degree Fahrenheit.

K-factor – Thermal conductance: rate at which a material conducts heat measured in BTU/in./hr.-sq. ft.-°F. (High K-factor = poor insulation).

R-value – Thermal resistance: capacity of a material to resist heat flow, the reciprocal of conductance. (High R-value = good insulation).

Roofs and Walls

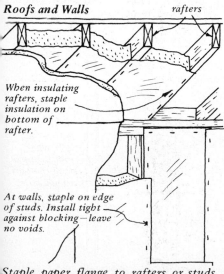

rafters

When insulating rafters, staple insulation on bottom of rafter.

At walls, staple on edge of studs. Install tight against blocking—leave no voids.

Staple paper flange to rafters or studs. Tape over joints in cold wet climates to prevent moisture in room from condensing in insulation.

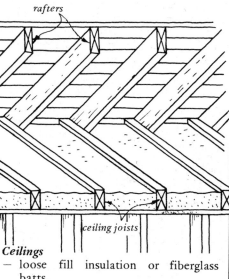

rafters

ceiling joists

Ceilings
— loose fill insulation or fiberglass batts.
— vapor barrier not needed in vented attics.

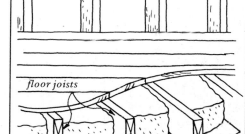

floor joists

waterproof sheet rock

Floors
Easiest is to staple chicken wire or nail waterproof sheetrock under joists, then lay fiberglass batts in from top *before* flooring goes on. However, in cold wet climates, a vapor barrier is advisable; this can be 15 lb. tar paper between subfloor and finish floor.

Application of Fiberglass or Loose Fill Insulation:
Note: always install with aluminum or vapor barrier on heated side, i.e., towards living space.

Insulating concrete slabs: see pp. 116-17.

Vapor Barrier

Warm air will hold more water vapor than cold air. Thus when warm moist air is cooled, it eventually reaches a temperature at which some of this moisture condenses into drops of water. The temperature at which this condensation occurs is called the *dew point*.

In uninsulated walls, the dew point is likely to be well out in the sheathing and condensation is not likely to occur within the stud space. In insulated walls, however, the dew point is brought in much further, and condensation inside the wall is more likely. Moisture destroys insulation value and can also rot the studs. It follows that if moisture can be prevented from escaping from the interior of the building into the wall, condensation is less likely.

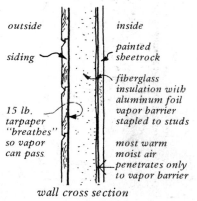

outside — inside

siding

15 lb. tarpaper "breathes" so vapor can pass

painted sheetrock

fiberglass insulation with aluminum foil vapor barrier stapled to studs

most warm moist air penetrates only to vapor barrier

wall cross section

A vapor barrier is an airtight barrier which prevents the warm moist air inside a house — due to cooking, bathing, people breathing, etc. — from reaching the inside of an insulated wall, roof or floor. In a wall or open ceiling it is placed over studs or rafters, close to the interior. In floors it is placed between subfloor and flooring. In vented attics it is not needed.

The aluminum foil on fiberglass insulation is an effective vapor barrier. Also used, in cold climates, is heavy, glossy-surfaced 50-lb. asphalt paper. Lighter weight tar paper is often used in warmer climates.

In practice a vapor barrier need not be totally airtight; however, the colder the climate, the more need there is to eliminate nail-holes and other leaks (by the use of tape at seams). Consult the building inspector for good local practice. Beware of foam-injected insulation in existing walls; it has been known to cause rot.

129

Sheetrock

Sheetrock (gypsum wall board) is relatively inexpensive: one-half to one-quarter the cost of wood paneling; it is fast to install, fire-resistant, sound absorbing, and can be textured and/or painted in many ways.

Installation:

— make big cuts by scoring with utility knife, then break board by bending away from scored side and slice paper on other side. Make small cuts (notches, electrical outlets, etc.) with keyhole saw.

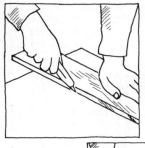

1. Score line on face paper side. Use straightedge.

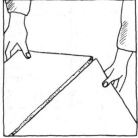

2. Break core by bending away from score.

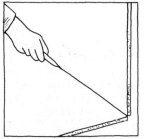

3. Slice through paper.

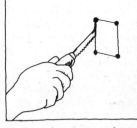

4. For switch-box drill ¼" holes at corners of outline, cut with key-hole saw.

— accuracy is not too important; loose cuts go up more easily.
— nail with cup-headed cement coated sheet rock nails.

Finishing:

— use pre-mixed joint compound (called "mud"); it comes in one and five gallon buckets.
— emphasis is on two-three *thin* coats; if too thick you must sand it off.
— fill large holes and cracks with quick drying plaster, such as Fixall.
— apply first coat of mud to nails and joints with wide knife. Use paper tape over joints, press into compound with knife, squeezing out and removing excess. (Thin coats!)

1. Nails every 7" o.c. - 3/8" inside edge.

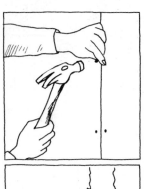

2. Joint compound applied to joints.

3. Press tape into mud full length of seam.

4. Sand. Then apply second coat, feathering out to smooth even finish.

— let dry thoroughly, sand lightly with No. 80 sandpaper, apply second coat to fill low spots.
— for smooth finish repeat above step again and sand lightly.

— for texture, you can use a stipple paint roller and roll mud on the wall. Or, for a more primitive effect you can spread a thin layer of mud or plaster over the surface with a cement finishing trowel.

Corners:

— tape inside corners (such as ceiling to wall) with tape that has been creased down the center.
— use metal corner trim on outside corners; then apply mud.

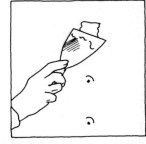

Use inside corner tool to smooth tape.

Use flexible putty knife to apply mud to nail holes in center of sheet rock.

Painting:

— any good interior wall paint; enamel for kitchen or bathroom, otherwise latex.
— two coats are usually necessary.

Alternate Way to Use Wallboard

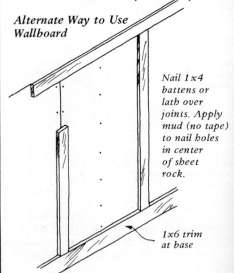

Nail 1x4 battens or lath over joints. Apply mud (no tape) to nail holes in center of sheet rock.

1x6 trim at base

Stairs

There are two important factors in stair design: all risers should be the same height, and all treads should be the same width. A handy formula for determining the rise/tread ratio is 2R + T=24 (R= rise, T=tread). This will work for stairs in any situation, from shallow garden stairs to a ladder. An average rise/tread ratio is a rise of 7, tread of 10.

Statistics compiled by the National Safety Council show that stairways are the largest single cause of accidents in the home. Below are a few details on stair construction. For full details on building stairs, see *Fundamentals of Carpentry — Practical Construction* (see bibliography).

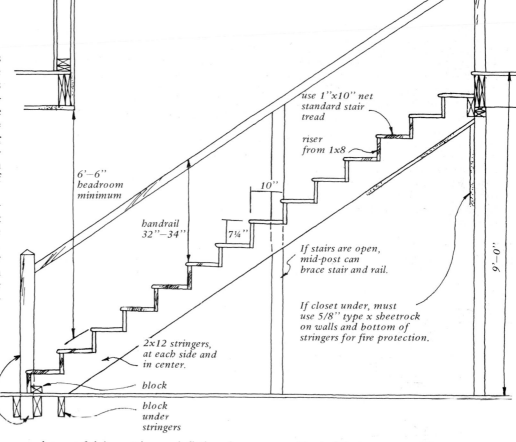

6'-6" headroom minimum

handrail 32"-34"

use 1"x10" net standard stair tread

riser from 1x8

10"

7¼"

9'-0"

If stairs are open, mid-post can brace stair and rail.

If closet under, must use 5/8" type x sheetrock on walls and bottom of stringers for fire protection.

2x12 stringers, at each side and in center.

block

block under stringers

4x4 newel post if stairs are open—seat between floor joists for stiffness.

Tips for Safe Stair Construction:
— use clear sound lumber.
— all treads and risers *must* be equal to avoid tripping.
— accuracy in layout is essential.
— drill for nailing to avoid splits.

— be careful in cutting and fitting the ends of the stringers to make sure that the first and last treads and risers are equal to all others.
— handrails are essential for safety and must be 32"-34" high.

Typical Stairs:
15 risers at 7¼", 14 treads at 10". Minimum width of stairs: 2'-6". Typical Building Code requirements: minimum 9" tread, maximum 8" riser.

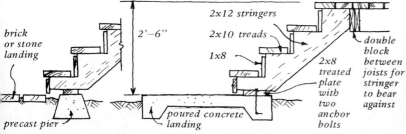

brick or stone landing

precast pier

2'-6"

2x12 stringers

2x10 treads

1x8

poured concrete landing

2x8 treated plate with two anchor bolts

double block between joists for stringer to bear against

Porch stair with precast pier

Porch stair with concrete landing

Porch Stairs:
— slightly steeper ratio of 7½" to 9½".
— bottom end of stringer rests on masonry landing; no earth/wood contact.
— good connection between stringer and header made possible by extending porch floor on to stringer to form top tread.

Plumbing

Water Supply

Choice of pipes: copper, galvanized steel, or plastic. Copper is quick and easy to install; galvanized steel is cheaper but more difficult to work with; plastic (only PVC or CPVC are allowed) is quick and easy to install, but allowance must be made for expansion when used for hot water. It is probably safer to use copper or steel for hot water.

Soldering copper pipe:

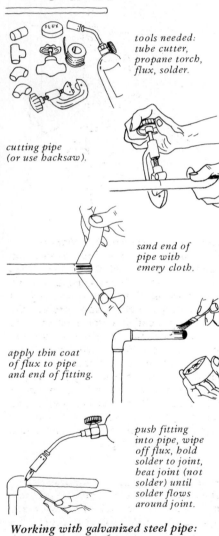

tools needed: tube cutter, propane torch, flux, solder.

cutting pipe (or use hacksaw).

sand end of pipe with emery cloth.

apply thin coat of flux to pipe and end of fitting.

push fitting into pipe, wipe off flux, hold solder to joint, heat joint (not solder) until solder flows around joint.

Working with galvanized steel pipe:

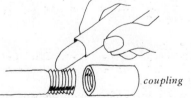

coupling

Apply joint compound, tighten joints with two pipe wrenches. Never back up or joint may leak.

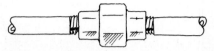

Pipe union used for cutting into and joining two pieces of pipe or where you may want an easy disconnect (such as at hot water heater).

Waste

What a trap does:

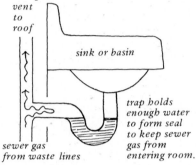

vent to roof

sink or basin

trap holds enough water to form seal to keep sewer gas from entering room.

sewer gas from waste lines

Traps and vents are necessary on all fixtures. Without a vent, rush of waste water will suck enough water from trap to allow sewer gas into room.

Use of ABS plastic drain pipe is much easier for the homebuilder than steel pipe. Pipe is cut with wood or hack saw and joined with solvent cement. It hardens immediately.

Simple plastic waste system:

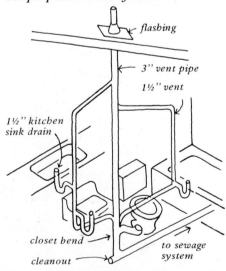

flashing

3" vent pipe

1½" vent

1½" kitchen sink drain

closet bend

cleanout

to sewage system

Note: individual fixtures may require separate vents if too far from stack.

Two good books on plumbing:
The Home Owner Handbook of Plumbing and heating by Richard Day (see bibliography) and *Plumbing*, Time-Life Books.

Electrical

Wiring

With electricity a little knowledge can be dangerous. But with a good book and help from an electrician, wiring can be done by the owner-builder. A very good book on wiring is *Wiring Simplified* by H. P. Richter.

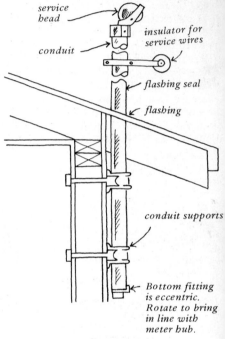

service head

conduit

insulator for service wires

flashing seal

flashing

conduit supports

Bottom fitting is eccentric. Rotate to bring in line with meter hub.

Service entrance

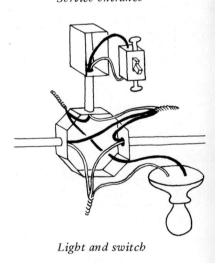

Light and switch

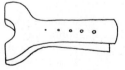

Romex cable stripper saves time and sliced thumbs, costs about $1.

Metal Chimneys

Safety

Metal Chimneys

Many fires have been caused by improper installation of metal flues for wood stoves or Franklin or Jotul type fireplaces. There are two important points: 1) a metal flue should never be run horizontally as the draft will not work well and ashes and creosote can collect and either block or cause collapse of the pipe. Manufacturers recommend an angle of no more than 30° from the vertical; 2) special pipe, such as "Metalbestos," or triple walled, or ceramic pipe, all with adequate air spacing, must be used where the flue passes through the ceiling and or roof.

At plumbing or sheet metal suppliers, ask for brochure describing metal chimneys. Diagram below is of a typical installation.

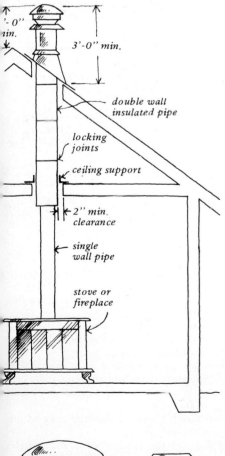

double wall insulated pipe

locking joints

ceiling support

2" min. clearance

single wall pipe

stove or fireplace

3'-0" min.

'-0" min.

round top chimney cap   *flashing adjustable to roof*

The accident rate is relatively high in the building industry. With a few precautions, you can minimize the chances of being injured:

Eyes:
- safety goggles for grinding, table saws, high speed machinery.
- safety spectacles are under $7, for people who do not wear glasses.
- plastic prescription lenses for builders who wear glasses.
- if you get something in your eye, flush with running water.

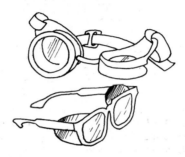

Ears:
- many older carpenters are partially deaf. Loud noises cause permanent damage.
- hearing protectors are best, ear inserts are better than nothing.
- ear protection should be used around power saws or other loud machinery.

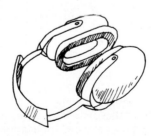

Skin:
- concrete eats away skin. Put heavy cream on hands (like Nivea) before pour. Rinse thoroughly after pour, put on hand lotion.
- rubber gloves for liquid hazards; leather gloves for wrecking, moving lumber, etc.

Lungs:
- disposable masks for non-toxic particles are cheap, easy to wear.
- single or double cartridge respirators for hazardous vapors or particles such as paint spray, fiberglass particles, fine wood dust, etc.; different cartridges for different hazards.

Head:
- lightweight yet strong hard hats should be worn around any heavy construction, or when working under anything that might fall. Can prevent painful bumps, or more serious injury.

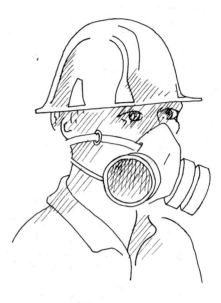

Home remedy for smashed thumb:
Soak in hot water as long as you can stand it, then in bowl of ice water. Then back into hot. The expansion-contraction loosens skin, helps swelling. If thumb keeps swelling and throbs, take paper clip, bend it open, heat one end until red hot, push into nail about 1/8" above nail bed. Push until liquid flows out: immediate relief. Squeeze as much liquid out through hole as you can, put on good germicide, bandage. In day or two if more fluid, sterilize needle and gently open hole so more liquid can escape. It may save the nail, and you'll be able to sleep at night.

Muscles:
Most back and muscle injuries occur when muscles are tense and cold. Stretching exercises can help prevent many injuries. A good book: *Stretching*, by Bob Anderson, P.O. Box 2734, Fullerton, Ca. 92633 will tell you how to keep all muscles flexible.

Safety equipment:
- hearing protectors: David Clark Co., Inc., 360 Franklin St., Worcester, Mass. 01613.
- goggles, spectacles, boots, gloves, respirators, etc.: Direct Safety Co., 511 Osage, Kansas City, Ks. 66110.

Nailing

Recommended Nailing for a Wood-frame House

Joining	Nailing method	Number of nails	Nail size	Nail placement
Header to joist	End-nail	3	16d	
Joist to sill or girder	Toenail	2-3	8d	
Header & stringer joist to sill	Toenail		10d	16" on center
Subfloor boards, 1 x 6		2	8d	To each joist
Subfloor, plywood:				
At edges			8d	6" on center
At intermediate joists			8d	8" on center
Bottom plate horiz. assembly	End-nail	2	16d	At each stud
Top plate to stud	End-nail	2	16d	
Stud to bottom plate	Toenail	4	8d	
Upper top plate to lower top plate	Face-nail		16d	
Ceiling joist to top wall plates	Toenail	3	8d	
Rafter to top plate	Toenail	2	8d	
Rafter to ceiling joist	Face-nail	5	10d	
Ridge board to rafter	End-nail	3	10d	
Collar beam to rafter:				
2-inch member	Face-nail	2	12d	
1-inch member	Face-nail	3	8d	
Wall sheathing:				
1 x 8 or less, horiz.	Face-nail	2	8d	At each stud
1 x 6 or greater, diag.	Face-nail	3	8d	At each stud
Plywood wall sheathing:				
3/8" and less thick	Face-nail		6d	6" edge
½" and over thick	Face-nail		8d	12" intermed.
Roof sheathing, 4,6,8" width	Face-nail	2	8d	At each rafter
Roof sheathing plywood:				
3/8" and less thick	Face-nail		6d	6" edge
½" and over thick	Face-nail		8d	12" intermed.

16d common, 3½" long.

16d box, 3½" long.

16d cement coated sinker, 3" long — for working with 1½" lumber.

8d common, 2¾" long.

8d box, 2¾" long — use hot dip galvanized for exposed nailing.

16d finish.

8d finish.

Notes:
1. *d refers to pennyweight. 6d means a "six-penny" nail.*
2. *There are two types of nails: common and box. Box nails are thinner. All nail sizes referred to above are common nails.*
3. *The 8d common and the 16d common are the most common nails used in constructing a building. It's a good idea to buy a 50 lb. box of each when starting any project.*

Glossary

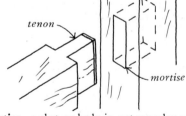

tenon

mortise

Abbreviated Glossary of Building Terms

birdsmouth: cutout near bottom of rafter which seats on wall plate.

cross tie: member used to tie two roof rafters together, as in gable roof, to make structural triangle.

d: refers to pennyweight of nails; 16d = 16 penny, etc.

girder: large horizontal member used to support joists or beams.

grout: thin concrete mixture used to fill voids, such as interior of concrete blocks.

header: top framing members over window or door opening; also, joists at ends of opening in floor used to support side members.

jamb: side and top members of window or door frame.

joist: floor framing member.

mortise: a slot or hole in cut wood member to receive tenon.

mudsill: rot-resistant wood member attached to concrete foundation wall; floor joists rest on top of it.

o.c.: on center

plate: top or bottom horizontal frame member of stud wall.

plumb: vertical; to adjust into vertical position.

rabbet: groove cut in board to receive another member.

rebar: steel bars for reinforcing concrete.

ridge board: top horizontal member of roof framed with rafters.

riser: vertical board at edge of stairway step.

screed: grade level forms set at desired height so concrete can be roughly leveled by drawing straightedge over surface.

shake: handsplit wood shingle, usually sawed on one side.

shim: thin piece of wood (like shingle) used to fill a gap.

shiplap: horizontal siding rabbeted along each edge to provide close joint by fitting two pieces together.

stringer: side member of stairway that supports risers and treads.

stud: vertical structural member of wall in frame building.

threshold: piece of material over which door swings.

t & g: tongue and groove, as with tongue and groove flooring. Tongue of one board fits into groove of another.

tread: horizontal board in stairway.

vapor barrier: membrane used to prevent passage of moisture through walls, floors or ceilings.

Working

by Studs Terkel

From Studs Terkel's interview with Nick Lindsay, 44 years old, father of ten children, carpenter-poet in Goshen, Indiana:

. . . Lindsays have been carpenters from right on back to 1755. Every once in a while, one of 'em'll shoot off and be a doctor or a preacher or something. Generally they've been carpenter-preachers, carpenter-farmers, carpenter-storekeepers, carpenters right on. A man, if he describes himself, will use a verb. What you do, that's what you are. I would say I'm a carpenter. . . .

Who are you working for? If you're going to eat, you are working for the man who pays you some kind of wage. That won't be a poor man. The man who's got a big family and who's needing a house, you're not building a house for him. The only man you're working for is the man who could get along without it. You're putting a roof on the man who's got enough to pay your wage.

You see over yonder, shack need a roof. Over here you're building a sixty-thousand-dollar house for a man who maybe doesn't have any children. He's not hurting and it doesn't mean much. It's a prestige house. . . .It's a real pleasure to work on it, don't get me wrong. Using your hand is just a delight in the paneling, in the good woods. It smells good and they shape well with the plane. Those woods are filled with the whole creative mystery of things. Each wood has its own spirit. Driving nails, yeah, your spirit will break against that. . . .

One nice thing about the crafts. You work two hours at a time. There's a ritual to it. It's break time. Then two hours more and it's dinner time. All those are very good times. Ten minutes is a pretty short time, but it's good not to push too hard. All of a sudden it comes up break time, just like a friend knocking at the door that's unexpected. It's a time of swapping tales. What you're really doing is setting the stage for your work.

A craftsman's life is nothin' but compromise. Look at your tile here. That's craftsman's work, not art work. Craftsmanship demands that you work repeating a pattern to very close tolerances. You're laying this tile here within a sixteenth. It ought to be within a sixty-fourth of a true ninety degree angle. Theoretically it should be perfect. It shouldn't be any sixty-fourth, it should be 00 tolerance. Just altogether straight on, see? Do we ever do it? No. Look at that parquet stuff you got around here. It's pretty, but those corners. The man has compromised. He said that'll have to do.

They just kind of hustle you a bit. The compromise with the material that's going on all the time. That makes for a lot of headache and grief. Like lately, we finished a house. Well, it's not yet done. Cedar siding, that's material that's got knots in it. That's part of the charm. But it's a real headache if the knots fall out. You hit one of those boards with your hammer sometime and it turns into a piece of Swiss cheese. So you're gonna drill those knots, a million knots back in. (Laughs.) It's sweet smelling wood. You've got a six-foot piece of a ten-foot board. Throwing away four feet of that fancy wood? Whatcha gonna do with that four feet? A splice, scuff it, try to make an invisible joint, and use it? Yes or no? You compromise with the material. Save it? Burn it? It's in your mind all the time. Oh sure, the wood is sacred. It took a long time to grow that. It's like a blood sacrifice. It's consummation. That wood is not going to go anywhere else after that.

When I started in, it was like European carpentering. But now, all that's pretty well on the run. You make your joints simply, you get prehung doors, you have machine-fitted cabinet work, and you build your house to fit these factory-produced units. The change has been toward quickness. An ordinary American can buy himself some kind of a house because we can build it cheap. So again, your heart is torn. It's good and not so good.

Sometimes it has to do with how much wage he's getting. The more wage he's getting, the more skill he can exercise. You're gonna hire me? I'm gonna hang your door. Suppose you pay me five dollars an hour. I'm gonna have to hang that door fast. 'Cause if I don't hang that door fast, you're gonna run out of money before I get it hung. No man can hurry and hang it right.

I don't think there's less pride in craftsmanship. I don't know about pride. Do you take pride in embracing a woman? You don't take pride in that. You take delight in it. There may be less delight. If you can build a house cheap and really get it to a man that needs it, that's kind of a social satisfaction for you. At the same time, you wish you could have done a fancier job, a more unique kind of job.

But every once in a while there's stuff that comes in on you. All of a sudden something falls into place. Suppose you're driving an eight-penny galvanized finishing nail into this siding. Your whole universe is rolled onto the head of that nail. Each lick is sufficient to justify your life. You say, "Okay, I'm not trying to get this nail out of the way so I can get onto something important. There's nothing more important. It's right there." And it goes — pow! It's not getting that nail in that's in your mind. It's hitting it — hitting it square, hitting it straight. Getting it now. That one lick.

If you see a carpenter that's alive to his work, you'll notice that about the way he hits a nail. He's not going (imitates machine gun rat-tat-tat-tat) — trying to get the nails down and out of the way so he can hurry up and get another one. Although he may be working fast, each lick is like a separate person that he's hitting with his hammer. It's like as though there's a separate friend of his that one moment. And when he gets out of it, here comes another one. Unique, all by itself. Pow! But you gotta stop before you get that nail in, you know? That's fine work. Hold the hammer back, and just that last lick, don't hit it with your hammer, hit it with a punch so you won't leave a hammer mark. Rhythm. . . .

From Working: People Talk About What They Do All Day And How They Feel About What They Do, *by Studs Terkel.*

High School Carpenters

LOW RENT HOUSING
JOINT EFFORT
HOUSING AUTHORITY

WITH DESIGN AND CONSTRUCTION BY
**SANTA BARBARA HIGH SCHOOL DISTRICT
CONSTRUCTION TECHNOLOGY CLASSES.**
Lender : Imperial Savings and Loan Assn.

Over the past six years, students in the Santa Barbara, Calif. public high school system have built four houses as part of an industrial arts program in home construction and repair.

For each project, local architects help in setting up a design contest. Students go to the architects' offices to learn drafting and design skills, often in the evenings. After student designs and models have been judged, one set of plans is selected for the current project. Students then take the plans through the planning and building inspection departments.

Over 80 students are involved in the program, which is coordinated by Mark Phelps. Local carpenters, roofers, masons, plumbers, electricians and painters serve on an advisory committee which provides on-site guidance.

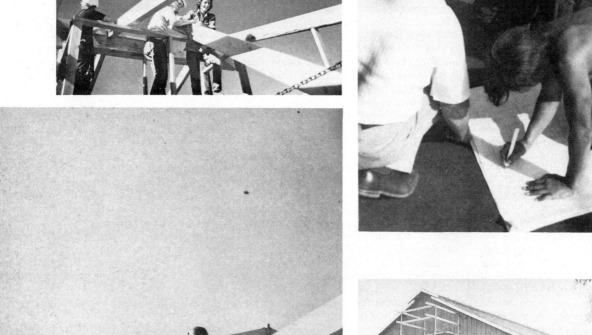

The teacher of the interior design course and her class select room color schemes, wallpaper, drapes, carpeting, sink tiles, lighting fixtures, etc. A horticulture teacher and her students landscape the grounds around the houses.

Once a house is finished, it is put on the market by the teacher and students in the real estate class. On two of the houses, the school district broke even; on the third, they made $7500 profit; the fourth has not yet been sold.

The district has an 18 acre farm 15 miles from the city and students are currently building a solar greenhouse and remodeling a barn. □

The next generation

Plywood bracing is very effective in strengthening a house against vibrations caused by an earthquake. This house in Anchorage, Alaska was ruined by a landslide caused by the 1964 earthquake. However, the plywood walls retained their shape, and prevented a complete collapse.

Earthquake

by Peter Yanev

The consensus among engineers who have studied the effect of past strong earthquakes on single family residences throughout the United States is that architecturally simple, single story wood frame houses generally survive without heavy damage, particularly when there are no failures of the foundation due to the geology of the site, i.e., landsliding, liquefaction, faulting. The damaged buildings are usually older, multi-level, and masonry houses.

Wood Frame Houses

A carefully designed and constructed modern wood frame building is, by far, the most desirable small property investment in earthquake country. The high earthquake resistance of wood buildings is primarily the result of the lightness and flexibility of the material.

A wood frame building is most likely to suffer serious earthquake damage

Peter Yanev is a civil engineer and author of the book Peace of Mind in Earthquake Country *(see bibliography).*

when one or more of the following conditions are present:
— a site underlain by soft, unstable ground.
— a weak or inadequately connected foundation.
— structurally weak architectural features (all glass and no walls, for example.
— an old, poorly maintained building.
— insufficient lateral (earthquake) bracing or an inadequate number of load-bearing walls and columns.
— improper placement or inadequate connections of lateral bracing and other structural detailing.
— slender, thin vertical supports, such as the stilts used for some hillside homes.
— heavy roofs, such as clay tile.

The most important condition for a durable and safe wood frame building is the lateral bracing, or earthquake resistant system, now generally required for all new construction in California. Diagonal bracing with 1 inch x 4 inch lumber has been the standard. Except for

single-story homes, that bracing is insufficient for California. Shear-wall bracing — a continuous covering of plywood paneling over the vertical framing studs — is recommended in lieu of diagonal bracing in areas subject to strong earthquakes and in all split-level or multi-story houses. Like any other bracing system, shear-wall bracing with plywood panels will be effective only if the wood is of good quality, and the nailing is adequate (see p. 134).

When applying the plywood directly to the studs, it should be at least 3/8 inch thick and, whenever possible, the panels should extend the height of the wall. The conventional "double wall" construction of sheetrock and stucco, which is very common in the western states, is not strong enough for high earthquake intensity areas, and such construction should be reinforced with additional shear-walls of plywood.

Older Wood Frame House Foundations

Old buildings tend to suffer disproportionate earthquake damage either because the construction practices or codes were inadequate, or because age, poor maintenance, or geological foundation problems have weakened the structure. The minor Santa Rosa, Calif. earthquake in 1969 provided some pertinent data on the relationship between the age of a building and its susceptibility to damage during an earthquake. Of 38 wood frame dwellings badly damaged in the tremor, 29 had been built before 1920 and nine pre-dated 1940. Wood frame structures built after 1940 sustained very little damage. The buyer or owner of an old building should also consider that earthquake insurance rates for such buildings may sometimes be more expensive.

Inadequate Foundation Connection

One of the most common failures of wood frame buildings results from insufficient or poor anchorages between the horizontal wood sill (often called the mud sill) and the underlying concrete foundation.

One of several buildings thrown off their foundations by the Inangahua, New Zealand earthquake of 1968. The building was not bolted down to its foundations

Few buildings built before 1940 have any connections between the sill and the foundation, except in areas that have been repaired for termite or wood rot damage. The lack of this detail is probably the single most important cause of earthquake damage to single story houses. Fortunately, existing wood frame buildings lacking sufficient (or any) sill connections can be rather easily repaired with the addition of ½ inch or larger steel expansion bolts wherever there is access to the sill. The ends of the sills are especially important and should be anchored even if it is necessary to temporarily remove the wall covering or bracing in order to add the bolts.

Typical damage to furnishings and kitchen wares from the San Fernando, California earthquake of 1971.

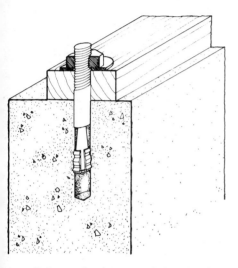

Detail for anchoring an existing foundation with expansion bolts. The bolts should be placed at a six foot spacing along the periphery of the house.

Cripple Stud Foundations
The typical cripple stud connections between the sill and the floor are another very common cause of earthquake damage in wood frame buildings. Cripple

Unbraced cripple stud walls are a frequent cause of building failure during earthquakes. Shown here is a building in the 1906 earthquake in San Francisco.

studs are short, vertical 2 x 4 supports which lift the floor one foot or more from the foundation and provide a crawl and air space under the building. Many serious failures during the 1933 Long Beach earthquake and among the new homes in San Fernando in 1971 are attributed to cripple studs which were braced, if at all, by a few unreliable diagonals. The solution to this type of foundation problem is quite simple — merely add plywood sheathing in lieu of or to supplement the diagonal bracing. The plywood sheathing provides a continuous surface which quickly transfers the earthquake forces to the ground while greatly strengthening the resistance of the cripple studs to the lateral earthquake motions. The plywood shearwall bracing should be fitted as tightly as possible against the cripple studs and should be thoroughly nailed to the sill, studs and floor joists or plates.

Masonry Houses
Pre-1940, unreinforced masonry buildings are the most dangerous of all buildings in California. Most deaths in past California earthquakes were caused by collapse of brick, stone, and adobe buildings. The reason for this vulnerability is simple — the masonry is heavy and fragile, and the strength of the walls depends on good mortar and workmanship to hold the bricks together. All this makes for potential disaster, as it did in the 1868 quake in Hayward; 1892 in the Owens Valley; 1906 in San Francisco, San Jose, Santa Rosa and elsewhere in the affected areas; 1925 in Santa Barbara; 1933 in Long Beach; 1952 in Kern County and Bakersfield, and 1971 in San Fernando Valley (all in California). In the San Francisco Bay Area, there are perhaps as many as 20,000 unsafe older masonry buildings. A similar number of such buildings can also be found in the Los Angeles area. These are the buildings that kill.

In general old, unreinforced masonry buildings are difficult and expensive to repair and reinforce. The cost of such work may exceed 50% of the value of the building. The idea that older buildings are substantial because they have stood the test of time can be a dangerous fallacy for earthquake country. The prospective buyer of an unreinforced brick building should consult a civil or a structural engineer, who can determine the feasibility and costs of making the structure safer. It is more desirable to pay the moderate fee for an engineer's consultation services, than to risk the entire investment and your life in a future earthquake.

Conclusions
A few basic factors govern the extent of damage to buildings during earthquakes. The most fundamental factor is the stability of land upon which the building stands or will stand. Anyone living in California and/or investing in California real estate should make the effort to determine the geologic conditions below the building site, the proximity of active faults and landslide areas, and the historic behavior of the land during previous tremors. Another important factor is the structural design of the building. Many structural deficiencies can be resolved readily, and architectural problems can either be avoided or corrected through reconstruction or reinforcement procedures. And, finally, there are the ultimate safeguards of earthquake insurance and intelligent behavior before, during, and after an earthquake. Surely, a thorough understanding of these basic principles, facts, and remedies, is a prerequisite for safety, sound real estate investment, and peace of mind in earthquake country. □

Materials

Basic Adobe
by P. G. McHenry

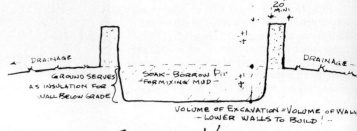

SHELTER WALLS

Depending on climate, walls can be made by:

Dry climate: Adobe bricks, molded by you and dried in the sun, or puddled shaping of moist earth by hand into as high a mass as will stand by itself, allowing to dry until firm, and adding additional layers until the wall height is achieved.

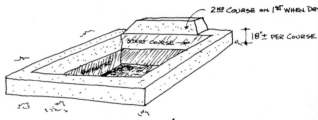

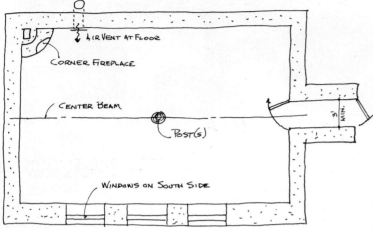

SHELTER PLAN

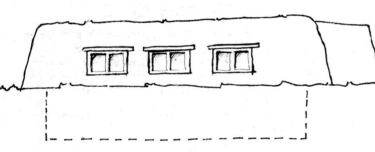

SOUTH ELEVATION

P. G. McHenry is a longtime designer-builder and adobe specialist. His classes at the University of New Mexico and the College of Santa Fe have resulted in over 50 homes and major projects by owner-builders. He is author of Adobe-Build It Yourself, *University of Arizona Press, Box 3398, Tucson, Ariz. 85722.*

Note: *load-bearing adobe walls and earth roofs can be highly hazardous in earthquake zones. Check local building codes and/or talk to local builders if you are in an area where earthquakes might occur.*

Comfortable shelter, in its simplest form, can be easily and cheaply achieved by using the same rudimentary principles our forebears employed. The basic principles have been in use for thousands of years, all over the world. The early Hohokam Indians of the southwestern United States found the basic form so satisfactory that they didn't change it substantially for almost 600 years. Can we say that of our "shelters" today? The following basic "rules" could be put to good use.

Rule 1. Choose a site that won't be flooded, and is well drained from average rainstorms.

Rule 2. Choose materials for construction that are close at hand. Earth is often the closest, as it is underfoot. The "labor factor" is the major owner-absorbable item, so we don't care if the labor is high in relationship to the material cost. If the material is "free" too, so much the better.

PUDDLED ADOBE

Damp climate: Use small forms (48" long x 18" high, plus or minus), tamping in damp (not wet) earth until it sounds solid (rings). Remove form, reset in another location and continue. Use rope, wire or even cloth ties to keep the form from spreading.

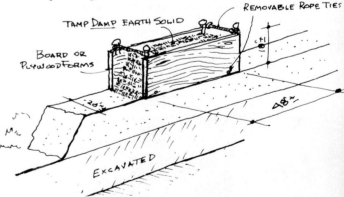

RAMMED EARTH

Note: The process can be speeded up and strengthened by setting palisade posts close together, tied and plastered with mud. (Cedar, redwood, cypress, aspen or rot-resistant woods are preferred.)

Soils: Almost any earth will do that seems to have a slight clay and sand content. If the soil has too much clay, sand or straw can be added to make a workable mix. Heavy rainstorms seldom penetrate more than a fraction of an inch in medium clay soils.

Rule 3. The key to safe, economical roof structures is a short span.

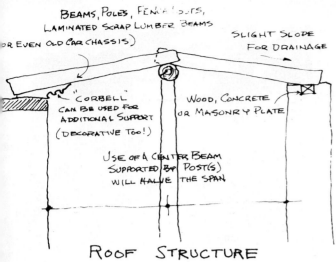

BEAMS, POLES, FENCE POSTS,
LAMINATED SCRAP LUMBER BEAMS
(OR EVEN OLD CAR CHASSIS)

SLIGHT SLOPE
FOR DRAINAGE

"CORBELL"
CAN BE USED FOR
ADDITIONAL SUPPORT
(DECORATIVE TOO!)

WOOD, CONCRETE
OR MASONRY PLATE

USE OF A CENTER BEAM
SUPPORTED BY POST(S)
WILL HALVE THE SPAN

ROOF STRUCTURE

Rule 4. *An earth roof is practical, insulative, cheap and easily repairable.*

The deck to support it should be made of rot-resistant wood if possible, but this is not required if you use a tar paper barrier.

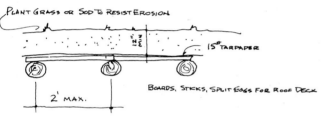

PLANT GRASS OR SOD TO RESIST EROSION

15" TARPAPER

2' MAX.

BOARDS, STICKS, SPLIT LOGS FOR ROOF DECK

NOTE: IF ROUGH DECK MATERIAL IS USED CEDAR BARK, ASPEN SHAVINGS OR MOSS SHOULD BE PLACED IN CRACKS TO PREVENT TEARING OF TARPAPER.

EARTH ROOF

Rule 5. *The shape of the building should be roughly rectangular*, but can be curved or rounded to make use of available materials. An "L" shape open to the southeast offers a more complicated but improved form, providing wind protection and a heat trap for the sun's warmth. A vestibule entrance and a minimum number of exterior doors will reduce heat loss. Entrance should be on the opposite side from prevailing winds.

Rule 6. *Place windows on south side* (northern hemisphere) to get maximum solar gain during the winter months.

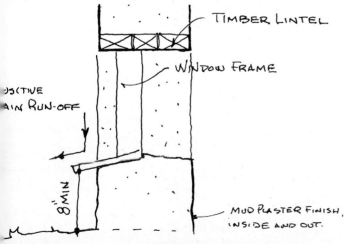

TIMBER LINTEL

WINDOW FRAME

POSITIVE RAIN RUN-OFF

8" MIN

MUD PLASTER FINISH INSIDE AND OUT.

PROVIDE INSULATION FOR WINDOWS AT NIGHT
– OR HEAT GAIN WILL BE LOST –

WINDOWS

Provide some form of insulation for windows at night or the heat gain during the day will be lost at night. Windows can be salvaged or job-built frames; double glazed is better.

Rule 7. *Provide heat with a corner fire pit/fireplace-stove* on the north side with a masonry flue. Maximum heat is accomplished from reflection from the back walls of the firebox. Flue size should be approximately 10% of the firebox opening. A combustion air vent and deflector wall placed close to the fireplace will reduce smoking, and provide summer ventilation.

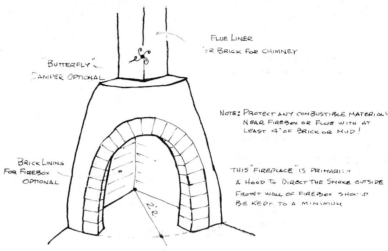

FLUE LINER
OR BRICK FOR CHIMNEY

BUTTERFLY
DAMPER OPTIONAL

NOTE: PROTECT ANY COMBUSTIBLE MATERIALS NEAR FIREBOX OR FLUE WITH AT LEAST 4" OF BRICK OR MUD!

BRICK LINING
FOR FIREBOX
OPTIONAL

THIS "FIREPLACE" IS PRIMARILY A HOOD TO DIRECT THE SMOKE OUTSIDE FRONT WALL OF FIREBOX SHOULD BE KEPT TO A MINIMUM

CORNER FIREPLACE

Rule 8. *Building code should be consulted, if a building permit is to be obtained.* In many cases, from a practical standpoint, building codes and regulations are not as strictly enforced against owner built residences unless financing is to be involved, and as long as they do not affect public safety or infringe on neighbors' rights.

Rule 9. *Work out the details on paper, before you start!*

Further Refinements:
— Cloth ceilings, wall coverings.
— Flagstone, brick floors, below grade wall lining.
— Built-in benches, beds, etc.

This shelter can be built into the side of a hill (south) or against an existing building if proper supports and connections are provided. These accommodations are primitive, but can be quite adequate, warm and snug. If it is possible to build a more conventional and sophisticated shelter later, this one will provide an excellent root or wine cellar . . . . □

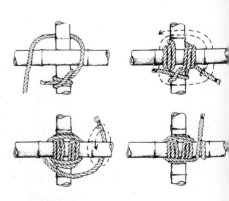

Bamboo

Por sus extraordinarias cualidades físicas, su forma y liviandad, el bambú ha sido el material de construcción de uso más diversificado que haya existido. Por su bajo costo y fácil disponibilidad, ha sido utilizado particularmente por la gente de pocos recursos económicos, tanto de latinóamérica como de algunos países asiáticos, que no solo lo emplean en todo tipo de construcción sino también en la elaboración de muebles y de infinidad de artículos de uso doméstico, por lo cual se le llama "la madera de los pobres". . . .

Bambu, *Oscar Hidalgo Lopez Arq. (vea bibliografia.)*

They rarely fail to impress and fascinate visitors to Hong Kong—the "spidermen" of the territory's construction industry.

They're the men who weave vast webs of scaffolding around building projects up to 52 stories.

What makes their work fascinating, even alarming, is the scaffolding itself. It's made entirely of bamboo, lashed together with thin bamboo strips.

To the visiting Western engineer whose own scaffolding has become a highly sophisticated science of steel tubing, supports, frames and bolted joints, the sight of a building swathed in a frail-looking latticework is often confused with technological backwardness — "Well, I guess they've been doing it that way for hundreds of years . . ."

In 1964 two 25-story buildings, one with bamboo scaffolding and the other clad with steel, were exposed to a typhoon which blew in with winds up to 120 miles an hour. The typhoon tore away the steel scaffolding, causing considerable damage, while the bamboo frame bent and swayed with the winds and withstood the storm.

Chronicle Foreign Service
August 14, 1977

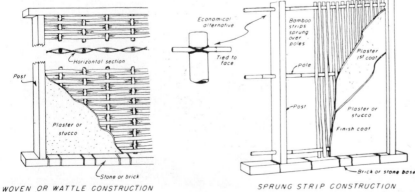

WOVEN OR WATTLE CONSTRUCTION   SPRUNG STRIP CONSTRUCTION

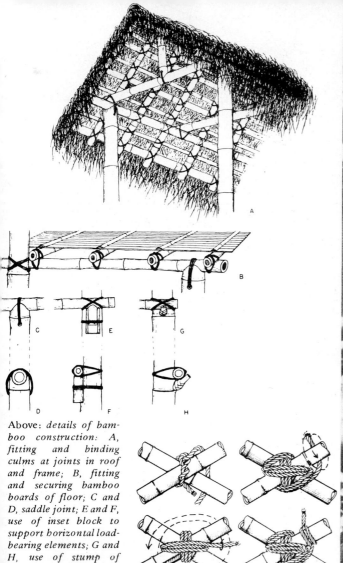

Above: *details of bamboo construction: A, fitting and binding culms at joints in roof and frame; B, fitting and securing bamboo boards of floor; C and D, saddle joint; E and F, use of inset block to support horizontal load-bearing elements; G and H, use of stump of branch at node of post to support horizontal load-bearing elements.*

A bamboo bridge over a river on Mount Khasi in the hills of northeast India.

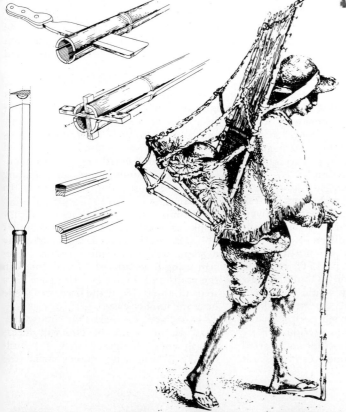

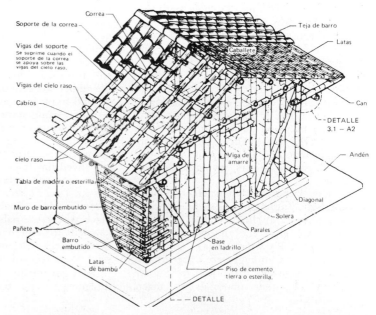

Correa

Soporte de la correa

Teja de barro

Vigas del soporte
Se suprime cuando el soporte de la correa se apoya sobre las vigas del cielo raso.

Latas

Vigas del cielo raso

Cabios

Caballete

Can

DETALLE
3.1 – A2

cielo raso

Viga de amarre

Andén

Tabla de madera o esterilla

Muro de barro embutido

Diagonal

Solera

Pañete

Parales

Barro embutido

Base en ladrillo

Latas de bambú

Piso de cemento tierra o esterilla.

DETALLE

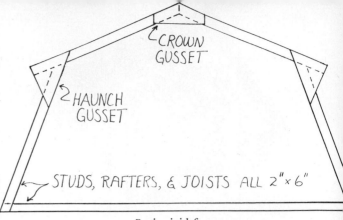

Basic rigid frame

The rigid frame cabin winter and summer.

Rigid Frame *by Ole Wik*

"Plywood rigid frame construction is the fastest, most economical way to enclose a space yet devised." So begins the 1969 booklet, *Plywood Rigid Frames*, published by the American Plywood Association. Having designed, built and lived for five years in my own rigid frame cabin, I can add that this kind of structure is very easy to build – and particularly enjoyable to live in.

The essence of rigid frame construction lies in fabricating a number of identical half-frames, each consisting of one stud, one rafter, and two specially-shaped plywood gussets (one on either side of the joint). These half-frames are nailed together at the peak with plywood crown gussets, giving a series of arches.

The arches are then erected on a suitable foundation, and plywood sheathing is applied to the roof and sides to give the necessary lateral rigidity. ("Plywood siding and roof sheathing are an engineered part of this design," cautions the booklet. "Substitution of other materials endangers structural soundness.")

Rigid frames provide unobstructed spans of up to 48', and the length of the building is limited only by the number of frames that one cares to make. Builders of barns, warehouses, aircraft hangars and the like have been quick to make use of this technique for rapidly and cheaply enclosing large spaces. Individuals planning smaller structures can also take advantage of the inherent economy and versatility of rigid frame construction.

I first came across rigid frame literature while doing research work on experimental housing for Alaska natives. Some years later, after having built and lived in three different semi-subterranean sod iglus in northwestern Asaska (see *Shelter*, p. 151), my wife Manya and I needed to build a new place where the baby would be born. I decided to combine the semi-subterranean, heat-efficient iglu idea with the rigid frame type of construction.

So I got out my booklet and started designing. Rigid frames can be built in either of two basic configurations: straight wall or slant wall. I chose the slant wall so that gravity would keep my insulation/heat bank of moss, sod and sand in place.

Next I had to decide on size. The booklet includes member selection tables indicating what size of framing lumber to use for various spans and roof loads, and also gives such fabrication details as the thickness of plywood to be used for gussets and sheathing, gusset dimensions for various spans, and nailing patterns for securing the gussets to the frames.

I found that the booklet didn't think as small as I did; spans under 24' were not included. So I did some extrapolating and came up with a design using 8' lengths of 2" x 6" for both the studs and the rafters. This gave me a span of about 18' at the base, and made ordering lumber easy.

I decided to make the house 16' long, to make use of two full sheets of plywood. I figured that if I built the back wall with duplex nails (the

144

kind with two heads), I could easily remove the entire rear wall at some later date in order to add more frames and extend the house as much as necessary.

The next question was whether my materials would hold up the weight of the sod and dirt that I planned to use as external insulation. I sent my design to the Association for comment, and one of the engineers in the Applied Research Department ran a mathematical check. He advised me that my materials would be seriously overstressed, and recommended shortening my frame spacing from 24'' to 16'' — which I did.

As for a foundation, my booklet indicated that rigid frames develop the outward thrust characteristic of all arches. This thrust is proportional to the weight of the building and to any snow load on the roof. In addition, there can be moderate inward and slight upward thrusts due to wind action on the walls.

I decided to use my floor joists to tie each arch together at the base, so that the thrust of one wall would cancel the thrust of the other. The joists exemplified the tight materials economy characteristic of the whole project: each 18' section was made up of two 8' and one 2' length

of 2'' x 6'', scabbed together with plywood scraps left over when the gussets were cut out.

When it came time to build, I dug a pit 2' deep in the well-drained sand at the edge of the bluff. I sawed two 16' logs in half, lengthwise, on a chain saw mill, and laid them down in the sand to support the frames. Once they were leveled, we stood the frames up, one by one. The thirteen arches and the wall and roof sheathing went up in a single day, and suddenly we had something that looked like a house.

Next I nailed the floor in place (using ¾'' tongue-in-groove plywood) and framed the end walls. All rigid frame buildings have vertical end walls that bear no load. Since my natural insulation would not stay in place against a vertical wall, I used 6'' of fiberglass instead.

Once 10'' of good moss covered the roof and the heat bank was in place, we were ready to go to work on the interior. Manya and I were pleased to find out how nice a "feel" the cabin had. Having lived for years in dwellings made out of natural logs and poles, we'd had misgivings about building with commercial lumber. Neither of us feels comfortable living in a "plywood box."

But this house is not that way at all, for two reasons: the walls slant, and the insulation is on the outside. The eye-pleasing arches formed by the exposed frames give the inside of the cabin the intriguing look of an upside-down boat. And as an entirely unexpected bonus, we found that we could build any number of shelves right into the walls, between the studs. This gives us instant storage space for every sort of household item, from books to bean sprouts, buttons to banjos. The lively swirl of shapes, textures and colors is just right against the steady, repetitive lines of the slanting frames.

As a builder, I am always pleased to hear Manya tell friends, "Of all the houses we've lived in, this one is the most fun." The arched ceiling is high enough for hanging laundry out of the way during the deep of winter, and we can go ahead and pound nails into the rafters any time we want to hang a cradle, jump-up or swing for the little people, or boot hangers, lanterns or tools for the grown-ups.

Our rigid frame cabin has done everything I expected it to do, and I look forward to building another some day. I will stick with the slanting wall and external insulation, but next time I'll use

rigid foam. My moss and dirt require maintenance every fall, since they tend to settle. This is especially troublesome along the out-sloping upper walls, since the slightest slumping there opens an air gap right next to the plywood skin. Then too, a moss roof is altogether too vulnerable to sparks. Next time I'll use proper roofing, and get a rainwater bonus as well.

Now that the suitability of the basic rigid frame cabin is taken for granted, I am thinking of several possible variations for next time, including the following:

— By adding a few external frames and prolonging the roof beyond the rear wall, one could construct a storm porch or woodshed that would be unheated but protected from rain and snow . . . .

— By extending the eaves, one could frame windows along all or part of the out-sloping portion of the upper wall, or else construct vertical bay windows. (These are standard rigid frame modifications.)

— How about using two sets of frames, of different sizes, in order to build a large main room with a smaller room of congruent shape attached at the rear? Or how about building two rigid frames next to each other, connected with a passageway? Maybe you can think of some other possibilities you'd like to try. Should you have any question about the structural soundness of your design, you might submit it to the Association and ask for a quick check.

If you ever do live in a rigid frame house, I venture the guess that you will soon develop an eye for other rigid frames that you never noticed before. Every time we go to Fairbanks, we pass a Carnation dairy building that has that characteristic, unmistakable out-sloping upper wall. □

For a list of plywood publications, write American Plywood Association, P.O. Box 2277, Tacoma, Wn. 98401.

Log Octagon

by Don Gesinger

An owner-designed, owner-built home: a specific case; an area northeast of Williams Lake, British Columbia with lots of timber, cold winters, and no enforced building codes. A natural 3-acre clearing between two creeks.

I wanted to build a house using native materials. My mate and myself did everything except for the neighborly helping hands. I bought as few building materials as possible. We lived in a tent at the site for a month *before* starting to build, during which time I worked out a plan in my head and decided on materials.

I used a mortarless stacked stone foundation, log walls with spaghnum moss insulation, pole floor joists and rafters, log trusses, and a hand-split shake roof.

Materials I had to buy were 2" x 6" T. & G. flooring, 1" x 6" roof boards, reflecting foil, used windows, ¼" safety plate for skylites, flashing, roof jacks, nails and caulking.

ROOF STRUCTURE

- SHAKES
- 1 x 3 STRAPPING
- REFLECTING FOIL, FOIL DOWN
- 2" x 4" AIR SPACE
- REFLECTING FOIL, FOIL UP
- 2 LAYERS OF 1" x 6" SPRUCE
- POLE RAFTERS @ 24" O.C.

MOSS LAID IN AS LOGS ARE FIT

T & G

POLE JOIST

CORNERSTONE BUILT-UP ROCK

Wall section

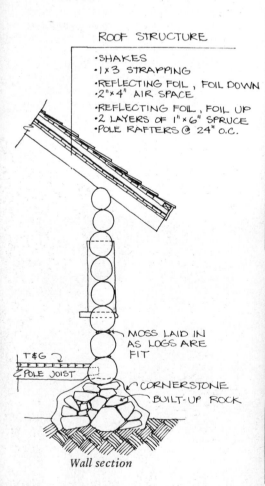

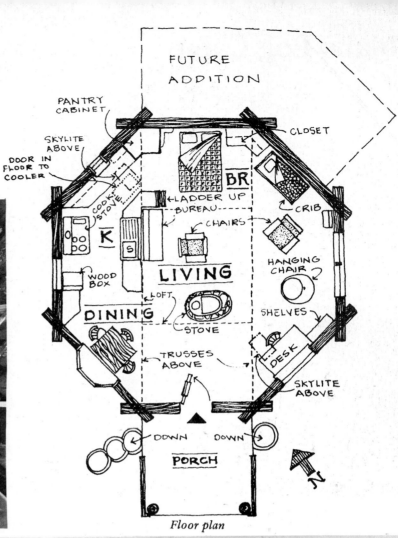

FUTURE ADDITION

PANTRY CABINET

SKYLITE ABOVE

DOOR IN FLOOR TO COOLER

CLOSET

COOK STOVE

LADDER UP
BUREAU

BR

CRIB

CHAIRS

K
S

HANGING CHAIR

WOOD BOX

LIVING

SHELVES

DINING

LOFT

STOVE

TRUSSES ABOVE

DESK

SKYLITE ABOVE

DOWN   DOWN

PORCH

N

Floor plan

The Plan: Logs are heavy and Jenny and I could only handle 14-footers. An octagon developed. I envisioned an unbroken polygonal space with more than one level: an octagon with a central loft (my studio) hung within the trusses that are part of the roof structure. I wanted to, and did, add a porch the second year; and soon will add two bedrooms. The result was a 650 sq. ft. house for under $700.00.

Focus: The Foundation. No cost. Partly the particular site was chosen because underground rock and gravel came to within 2" of the surface. Good bearing. I laid out the foundation walls 30" wide and removed the topsoil. Next, using a neighbor's front end loader, I moved eight large rocks to the eight corners and roughly leveled them. Next the first four opposing logs were set, crown up, and scribed level onto the eight rocks. The next four were fit to the first four, completing round one. I then built rocks *up to* the logs at all points between the eight cornerstones, installing a pre built crawl space access door frame at one point and laying stone around it. By the time the walls were finished they had settled down on the rocks so that a stone slipped into place could not be slipped out. □

Idaho Log Cabin

They paid $19 for a building permit, which they consider money well-spent since the inspector "... suggested several structural changes which strengthened our new home...." and thus they didn't have "...paranoid fears of being discovered and red-tagged during construction...."

Bill Namaste and Gina Gormley built this eight-sided log cabin in heavy snow country in northern Idaho. It cost $700, took a year to build, and "...was worth every penny and minute though, because our unique and beautiful new dwelling has kept us warm and dry during the worst of winter's storms ...." The 195 logs they used are from a neighbor's stand of lodgepole pine; they were flattened by running a chainsaw through them, then chinked with fiberglass insulation.

A large stone fireplace in the center is strengthened with reinforcing steel and provides flues for the heater and cookstove as well as support for roof rafters.

148

West Virginia Log Cabin
by Pete Lundell

It was back in the winter of 1973 that we found our hidden paradise — one hundred acres of beautifully secluded lush woodland. Miles to the east lived the old-timer from whom we purchased the property. He was quite a homesteader himself. As we explained why we wanted the land, his eyes seemed to glow with a vigor that he knew from days gone by, times when his grandfather carved out of a wilderness the present site of his home. The stories poured forth as we sat in the semi-dark room that evening. During those long pauses, all we heard was the crackling of the wood stove, it being the only light in the room.

"O.K., you can have it, $3,000 for all of it, the mineral rights and gas well t' boot." As he continued, he told us that he wanted us to build that log house on the very ridge that it now stands on. It was a place he once entertained in his own mind, but by now his youth had escaped him.

Backpacking the trail that led to the property found us on a high ridge, looking out from which there were miles of green forest fading into blue hills. This seemed to be the spot, intoxicatingly beautiful . . . .

Peter and Connie Lundell then went on to build this 36' x 56' log cabin on their 100 acre parcel of land in Hurricane, West Virginia. They cut all the pine logs on their property and hauled them to the building site with a four-wheel drive Jeep pick-up truck. The foundation is 40 large creosoted locust wood pillars, resting on stones deeply embedded in the ground; the stones, some weighing 500 pounds, were salvaged from an old log barn and an abandoned log cabin.

The fireplace is built of local stone and is used for cooking soups and stews. The floors are of oak odds and ends, fastened with walnut pegs.

They have had a half-acre lake built, which attracts deer, bear, bobcat and small game. They find, as did the log cabin poet and naturalist John Burroughs, that " . . . the most precious things of life are close at hand, without money and price."

Building with Stone

by Lewis and Sharon Watson

In 1973 my wife and I slip-formed an 1100 sq. ft., 3-bedroom, more-or-less conventional stone house in rural southern Idaho. Materials cost was about $2,000, building time about five months. We worked totally alone (except for "hand-up" help from our two girls, aged 9 and 11). Unfortunately, as school-teacher dropouts, we knew almost nothing about building (this was just before the current flood of owner-builder interest and information), so we made plenty of mistakes, mostly of the omission or missed-opportunity type. We put down everything we learned, positive and negative, in our book, *House of Stone*, which we self-published and have been selling by mail as a homestead income (*see bibliography*).

Based on four years of living intimately with our construction errors and successes, the following "tips" are a brief attempt to update and extend first impressions:

12 Tips for building and living in a Slip-formed Stone house

1. In slip-forming, make a double set of forms and use the Nearings' (*Living The Good Life*) "hand-over-hand" technique rather than our improvisation of single-set plus vertical supports. Form setting will go faster, you'll have greater continuity in stone-face finish, and probably better-cured concrete. Cost of extra forms is maybe $50, but results are worth it.

2. Slope top of each successive "pour" *downwards toward outside* to "shingle off" moisture that might otherwise soak through wall in probable absence of good bond between pours.

3. *Don't* pre-mix sand and gravel to save double hauling costs from gravel pit. Ratio of sand to aggregate gets screwed up and you get porous, gravelly and *leaky* walls. Custom-mix your own concrete batches, using lots of sand and trowel concrete solidly under, against, between, and behind *every* stone.

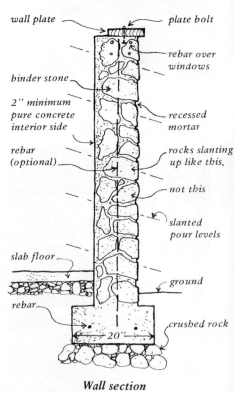

Wall section

wall plate — plate bolt — rebar over windows — binder stone — 2" minimum pure concrete interior side — rebar (optional) — recessed mortar — rocks slanting up like this, — not this — slanted pour levels — slab floor — ground — rebar — crushed rock — 20"

4. Unload your rocks in one big heap *inside* your intended house dimensions. That way you have to lug rocks shortest distance to all four walls. You also have optimum *selection* of rocks at all times, which is very important in fitting hundreds of odd-shaped stones to each other. This also means you probably shouldn't pour the slab floor until after the walls are complete, if you want to avoid chips and cracks.

5. Make absolutely sure you "key" window and door frames (plus top-wall plates) *into* concrete, i.e., insert a continuous strip of metal or wood into a groove cut on all sides of all frames before pouring concrete against frames. This "key" prevents serious drafts which would otherwise result as shrinking frames and concrete inevitably draw apart.

6. If tops of windows and doors can be finished flush with tops of walls, you can avoid building stone and concrete "bridges" there. If, for whatever reason, you wind up with rockwork over doors and windows, lay several rebars (or miniature "I-beams" made just for this purpose) in concrete and extend well on either side of opening. This provides support for roof trusses, etc.

7. Make the extra effort of planning *all* your interior finish work, paneling, electrical installations, etc. *in advance* of pouring stone walls. You can then anticipate needed tie-in or anchor points and insert attachments of whatever kind directly into the wet concrete. This is incomparably stronger than drilling holes in hard concrete later, using glue, etc.

8. Try to *finish* stone wall face as you go. We found pointing (troweling outside cracks between rocks with mortar) a messy, time-consuming job. We still haven't done so (we like the recessed-mortar appearance better), and no freeze-cracks or popped-off rocks have happened yet. Also, clean those innocent powdery concrete drippings from wall face as you go — they're nearly impossible to remove after several weeks of curing.

9. A stone house is *very* hard to "build onto" as new ideas and improvements occur to you in future years. You just can't casually knock out a wall for a new family room, extra bedroom, etc. For this reason alone, spend *lots* of time trying to anticipate future needs and impulses — then integrate those plans in your work, before the concrete sets.

10. You can save a lot of money, as we did, by buying doors, windows, fixtures, etc. at salvage yards, or maybe by tearing down one or more old houses yourself. Consider, though, that just maybe you'll actually want to sell your dream house later and really junky salvage materials and make-do building techniques will seriously drop the selling price. Upgrade your selection and performance just a bit and your stone house will be an excellent financial investment besides being a creation in which you take well-deserved pride.

11. Yes, you *can* — as a presumptuous and insecure amateur — build a pretty good house without knowing or using all that technical stuff tossed about and presumably used by construction professionals. However, it's a bit romantic to think that your very amateurishness and technical ignorance are *total* money-and-hassle-saving assets. Bite the bullet for a few months before beginning work by seriously studying all the good building books you can find. You may still choose to reject much of the overdone, over-slick, rigidly conventional and intolerably expensive techniques and materials of the pros, but it's amazing how much of it all you can usefully *adapt* to your own funky visions.

12. Lots of folks think a concrete slab floor is for the birds, cold, unattractive, and hard on the feet. We disagree. For looks and warmth, you can paint or tile your slab and/or cover it with throw rugs or carpet. And any good soft-soled shoes should meet the collapsing arches argument. On the positive side, we also like the way our slab cleans, its durability, and, of course, the initial simplicity and low cost of construction. □

Homes

154

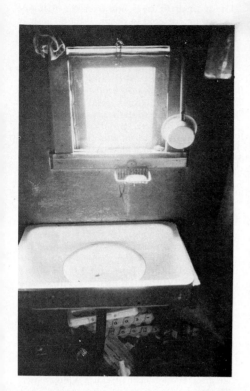

Orkney Islands

By Ruth Wheeler

Ruth and Jan Wheeler recently settled in the wind-swept Orkney Islands, off the northern coast of Scotland. The following article appeared in the Spring 1978 issue of the British magazine Undercurrents. *Since then, the Wheelers report, "We've covered quite a bit of ground . . . (we are) now killing our own meat (pigs and sheep), curing skins and spinning wool . . . . Also we've started on our second phase of building . . . " Ruth Wheeler is now working on a book that covers their first two years in the Orkneys and deals with " . . . the psychological difficulties of change-over from an urban to a rural lifestyle . . . ."*

Egilsay is a small island in the North Orkney group. It is only three miles by one mile and has no piped water or electricity. I came here with my husband and small son in 1974 from Brighton. We were the first English people to settle on the island and the native Orcadians sat back to watch our activities with barely disguised amusement. Here we were, three *ferryloupers* (incomers), with thirteen acres of uncultivated land, an old *but and ben* dwelling house and a few crumbling outbuildings.

My new "house" — I use the term loosely — was unbelievably primitive. Sheets of cardboard had been nailed on to the bare stone walls, and the former occupant had covered it with gay floral wallpaper. Rotting, sagging beams sighed under huge stone flags. Outside they were covered with *Orkney thatch*, that is, platted grasses and reeds held on with a great fishing-net. The floor consisted of large uneven flags laid directly on to the earth. On wet days mud oozed up between the chunks of rock.

The ravages of winter

The house had three rooms. Two were totally uninhabitable and the dimensions of the third main room, which I have described, were approximately thirty feet by twelve. Two huge box beds covered most of this room and an ancient Victoress stove stood in the massive fireplace.

We lived in this one room until the ravages of winter made it unbearable. The thatch needed replacing and the strong gales began to drive the rain in between the stones on the roof. Each night before turning off the oil lamps we placed buckets in strategic places and slept fitfully. Each morning we found a new spot where the water had managed to seep in.

We had done no building before and the thought of it made us feel very sick. We had so many other urgent tasks at hand. We had to prepare land, untouched for over twenty years, for the next growing season and we had to buy some livestock. We decided to leave the building until after the winter and we were fortunate to find a mobile home which we bought reasonably cheaply. Getting it to Egilsay was not so easy, of course, but we managed. This meant that we didn't have to rush the building and we were able to look around for good second-hand materials such as wood, roofing tiles, etc.

Undreamt-of luxury

By Christmas '74 we had brought electricity to "Whistlebrae" in the form of 1¾ kw Lister diesel engine; and on New Year's Day 1975 we got the water pump to work. For the first time in six months I didn't have to carry water from the well for ourselves and the livestock. I didn't know what to do with my spare time! . . .

Although, early in 1975, we had water on tap we were still heating it on the stove. Later we managed to get a second-hand back-boiler for the Rayburn and, with some copper pipes and a little ingenuity, we were able to have a real hot bath. (We use waste wood and drift-wood for the stove and so get all domestic hot water for free — not to mention the welcome blast of warmth in the winter months!) Anyone who has bathed in an old tin bath in a cold draughty room will know the joy of turning on the hot tap and stepping into a clean white bath. After living in such a primitive style we knew the real value of all resources.

Short, windy, winter days

I thought that August was a rotten time to come to a croft. What on earth could we do in the garden? — especially in a place like Orkney. But we did have one advantage. By the time the next growing season came around we had experienced an Orcadian winter and it was severe. From the end of September onwards we were constantly blasted by Westerly gales; and from mid-October until well into February darkness falls at about four in the afternoon, and it isn't really light enough for work until 8.30 in the morning. The gales drove the rain and hail with a hell of a force and with our 485 yds. of sea frontage, and a house situated very close to the shore, we were very exposed. We soon knew that our main problem would be one of shelter, and we spent that first winter building dry-stone walls. In biting winds we dragged huge boulders up from the beach and heaved them on to the walls. Without a tractor and trailer it was heavy and exhausting work.

I must stress here that we've had no tuition in this manner of building. However we have had a lot of practice and the shelter provided is second to none. Behind these walls we grow a variety of soft fruits, outdoor tomatoes and even sunflowers. Placed correctly, the stone holds the heat from the sun and this is an added advantage. This summer ('77) we've had good crops of blackberries, gooseberries, black-currants, and strawberries. We have ex-perimented with other shelters. We planted a lot of small fir trees but after the first gale they were battered to death. Trees do grow in Orkney but they all seem to be behind solid stone walls! The only other alternative shelter is of wood — that is, a barrier of wooden slats, with good foundations. They seem to be able to break the force of the wind.

Growing and storing the crops

To date we have grown good crops of 'nips (turnips). tatties (potatoes), carrots, onions. cabbage, cauliflower, sprouts, beets, parsnips, celery, lettuce. radish, etc. The greenhouse (well sheltered) provides us with tomatoes, cucumbers and the odd sweet corn. The surplus tomatoes are frozen, bottled and made into chutney. We also freeze a lot of peas and broad beans which grow well here. We run a large chest freezer off the generator, and although it is only on for a daily maximum of six hours in the winter, all food remains fresh and in good condition. So that we need not run the generator when there is almost perpetual daylight, in June and July, I empty the freezer. By mid-October it is full to capacity again with winter stocks.

We now grow over an acre of vegetables which last all the year around. What can be left in the ground that is, 'nips, carrots, beets, and parsnips, we leave. Potatoes are taken up any time between mid-October and mid-November and we store them in soil and straw. We grow enough cabbages of a winter variety (Drumhead), to keep ourselves and the livestock going through the winter. When the greenhouse is cleared of tomatoes at the end of October we dig over the soil and plant lettuces for winter salads. In fact, with careful planning and good shelter, one can have a wide and varied diet even in the toughest climate in the UK.

In November or December we bring cartloads of seaweed and shell sand from the shore and spread it across the growing areas. The contents of the pig, goat and hen houses also provide us with good manure.

Keeping the animals

Right now we have three milking goats, one milking cow and two Ayrshire calves — we favor Ayrshires because they are hardier than other milking breeds. We outwinter our animals, bringing them inside in only the severest weather. We supplement their diet from November until the beginning of May with hay, oats, 'nips, and cabbages. Most farmers here have beef cattle and bring them inside for five or six months, during the winter. We try not to buy any processed animal feed such as cattle nuts and the like. Whenever possible we feed bought-in calves on surplus goat's milk. If one cannot operate like this then the economy of the crofter or self-supporter would break down completely.

We also have sixteen maran hens, a sheep and a pig. The latter two will be killed and put in the deep freeze well before Christmas. The pig has been fattened mainly on fish. Jan fishes just off shore here for most of the year. He has a twelve foot dinghy and manages to catch good supplies of the local fish which is called sillock. We freeze the fish, then give the pig her daily quota mixed with potatoes, scraps, milk and whey from cheese making. We manage to get eggs throughout the winter by taking eight or nine selected hens into a larger henhouse where an electric light burns as long as the generator runs. They need more food, of course, but we have lots of tatties and also a few oats, and plenty of fish. Occasionally we have a spare hen who ends up in the pot.

Re-building the house
Last February we decided that we had had enough of living like gypsies and felt the need for four substantial stone walls around us. We had so many ideas. One was to buy a house in kit form and assemble it on a chosen site. We discarded this idea, mainly on the grounds of cost. The brochures talked glibly of thousands of pounds as if it was of no consequence. The idea of rebuilding the original dwelling house which was now in a sad state was depressing to say the least.

Then Jan had an idea which at first seemed impractical. We now had a 32 ft. mobile home which had proven its worth and strength over three Orcadian winters. In it was a lounge, bedroom, kitchen, bath and w.c. — in fact, all the plumbing and electrical wiring had been done. Why not rebuild the old but and ben house and put the mobile home alongside to form part of the house? Eventually we would roof the whole lot — house, mobile home and all. At first I disliked the idea. When we went into the cost of buying plumbing equipment for the house, bath, sink unit, toilet, piping, water tanks, etc., I realised that Jan's idea of utilising what we already had was the only practical one.

By March we had collected 1,600 second-hand Redland concrete roofing tiles and two hundred pounds worth of second-hand wood of varying thicknesses and lengths. In mid-March we started. We hauled the remaining thatch off the roof of the old dwelling house along with the massive flags. We then ripped away the rotting beams.

Extra northern lights
The walls were approx. two-three feet thick and we took three of the four walls almost to ground level in order to put in larger windows and doors. Orcadian windows are tiny. The natives of these islands think that windows make houses cold. We are gradually winning them over to the idea that light can also mean warmth. We got hold of three good window-frames with a job-lot of second-hand wood and they fitted well into the building. We put them on either side of the front door, retaining the original symmetry.

Because we didn't apply for a Local Authority grant we were able to decide upon the overall height of the building. We kept it low enough to retain its original character, but high enough to prevent severe brain damage from cracking our heads on the top of the doors! We built up the walls using the original stone along with some concrete. We gathered the sand from a spot well back from where the tide could reach, and when we had collected gravel from the shore we left it for a while to "wash." Once the window frames were in place we put up battens for the lintel above the front door and the large windows, and also at the sides for the gable ends. The front lintel was to be about 30 feet long. Over the frames we put steel rods and any other scrap metal available, in order to strengthen it.

Next we sorted through the wood and chose the strongest pieces for the beams. We had 18 lengths of pine which measured 18' x 5" x 3". We made a prototype in accordance with the angles we had already decided for the roof drop.

When the beams were fastened in place we put on the *sarking*. This was in fact the tongued-and-grooved flooring which we had bought with the second-hand wood. On top of the sarking we fixed tarred roofing felt, fastened securely with narrow vertical slats. Once we had a roof of sorts we could begin to work inside. It was September now and the weather was beginning to break.

Retaining the best features
We managed to retain the best features of the chimney-breast wall with its huge fireplace and spit, and the carefully placed stones which must have been there for several hundred years. These we scrubbed with a power drill and sander. We have left them completely naked. The other three walls are cement rendered and whitened, and we are told by the islanders that it looks very much like an old Orcadian house. We do not plan to have a ceiling as we feel that the natural wood of the huge beams and the sarking is far superior. Because the building is still fairly low there is not too much of a heating problem. We have an old Bonnybridge Dover stove with oven and copper water tank and this is adequate. If there is sub-zero weather we can light the fire in the large fireplace.

Two weeks ago we swung round the mobile home and our next job is to build a porch between the two. Then we shall start on the bottom section of the house following roughly the same process as we did for the main part. Eventually we plan to surround the mobile home with local stone so that it will really be part of the building. However this is going to take time.

We are pleased with what has been done so far. We have kept the long low look and costs have been kept to a minimum. Because Jan and I have done the work alone it has taken such a long time. The only help we have had was from one guy whom we hired for two

days when we hoisted up the beams, and another neighbour helped us for a few hours when we moved the mobile home.
Hard, heavy work
We found one of the most difficult jobs was the moving of the old flagstones on the floor. We dug up each one, placed soil and sand underneath in order to level it off. Finally we cemented round each one. We were going to have a cement floor but finally we decided that restoration work was more in order. However the large stones were very heavy and it took every ounce of our combined strength to move them. I have mixed concrete and heaved stones up and down ladders all summer.

We devised a pulley system when we had to take rocks up on to the roof for building up the chimneys. This consisted of two ladders and some strong rope. I tied the stones at the bottom of the ladder and Jan slowly dragged them up. Some of the work looked impossible. We've certainly learned the full extent of our strength, and we know our limitations. The work would have been easier with more bodies around, but as the community on Egilsay is a small and ageing one there isn't much chance of any practical help. In fact there are only 27 people on the island right now. There is a one-teacher school with six pupils and the population level has just about reached crisis point. Indeed, we are isolated with few services and the climate can be harsh, but it's a great place for adults and children.

Big is wasteful
Of course Government policy on agricultural land doesn't help remote islands like Egilsay. We haven't escaped the onslaught of the "big is beautiful" brigade. Farmers are encouraged by bribes (otherwise known as grants and subsidies) to swallow up smaller units and incorporate them into large agri-businesses. Houses are left to dereliction and decay. There are eight or nine unoccupied dwellings on this island in various stages of dilapidation which once belonged to small farms and crofts. Apart from our croft of thirteen acres there isn't a unit here under 50 acres.

We are trying to promote the reverse ideal of "small is beautiful." Indeed we grow a greater variety of food than anyone else on the island. We were told on arrival of the things which *wouldn't* grow. However with our shelters and planning we did what was thought to be impossible. Most of the natives had never tasted radishes or celery and they shook their heads in disbelief as we planted apple, pear, cherry and peach trees. As yet we've had no fruit as the trees are still too young, but we can say that they are alive and well and living in Orkney. □

Hoedad Yurts

by Charles Crawford

It all started in May of 1974. My friend Geno read the article in *Shelter* about yurts: he had to build one for himself. He was a Hoedad tree-planter here in Oregon and immediately saw the potential for portable structures to take into the woods. He built one with the help of his crew but it was of low quality and leaked terribly. No one was very impressed. Geno, however, was not one to be put off by a leaky roof. He immediately set to work planning another try and this time he came to my shop to work and collaborate, feeling that two heads are better than one.

Once I understood the basic concept, I was able to apply my knowledge of available low cost building technologies to the experiment. I had never seen a yurt up close before, so my first yurt exposure was when we put this one up, nearly a year later! We both became thoroughly engrossed in the project — improving on Geno's first yurt try and evolving the design to make better use of materials. We also spent some time trying to figure out how the Mongolians solved particular problems with their rude technology. My goatwool hat is off to them. In the end we turned out a nice little yurt that was

The Hoedads are a cooperative association of some 200 people in the Pacific Northwest engaged in reforestation.

very lightweight and structurally stable, but not strong enough to suit us.

A few months later, I started planting trees with Geno's crew. My search for a suitable living situation in the woods intensified my interest in building a light-weight portable structure capable of withstanding gale force winds, rain and cold throughout the many months spent living in the mountains. Most of the other crew members chose to live in rigs — house trucks, trailers, vans, and even station wagons. All of these alternatives seemed too expensive to me for the amount of comfort they provided. Everyone complained of not being able to stand up and move around without having to go outside. My search officially ended when Geno and I took the yurt we had built to our spring contract in the Wallowa mountains of Oregon. By the end of the contract, we decided that we should try to have a yurt, in which we could live year-round, built by the beginning of the new season. The resolve once taken was quickly acted upon: our first contract in the fall saw us living in a yurt that exceeded all of our expectations. Within the year, other members of the co-op were wanting to build yurts, but neither Geno nor myself had time to help them. Instead, I used my spare time to write a small manual.

As a result, virtually every crew in Hoedads has at least one yurt. Many of the crews have as many as five at a time being used as common space and individual home space. As skepticism fades, more and more people are showing interest in this sort of portable shelter.

In 1976, the materials for a 16' economy grade treeplanter's yurt cost $200. A 20' yurt cost $250 and a 25' yurt cost $285. If the materials were salvaged, the cost was considerably

158

less. Well over one hundred yurts have come directly of Geno's initial interest, and my efforts to make information available to as many people as possible. It would be hard to estimate how many yurts may actually have been built as a direct result of the article in *Shelter*!! The instruction booklet has grown from 16 pages and four illustrations, to 83 pages and numerous illustrations . . . .

Cascade Shelter came along as an aid to those who haven't the time to build a yurt, and as an easy way to gather the many and varied materials needed to build a yurt. It is a group of people banded together as a co-op to build yurts and to explore low cost shelter alternatives. The co-op provides flexibility by offering a variety of kits for all requirements and the ability to order individual parts or custom parts along with selling the completed product. CS does not build the traditional yurt as many may know it. The design is very practical — intended to be lightweight, strong and easily struck.

A manual for the more traditional design may be obtained through Chuck and Laurel Cox, Frog Pond Publishers, The Meeting School, Rindge, New Hampshire, 03461. This plan describes a 16' and 20' portable, insulated yurt made of either hardwoods or softwoods. It does not go into great detail. The Cox's plan includes a skylight. The frame is finely finished . . . . Cost is $4.00. I am sure that you are familiar with the Yurt Foundations, c/o William S. Coperthwaite, Bucks Harbor, Maine, 07618. Our manual, *How to Build a Portable Yurt* is available for $8.00 from Cascade Shelter, 4500 Aster St., Springfield, Oregon, 97477 . . . .

Our real interests lie far beyond yurts. They are a sensible place to start, but do not begin to solve the shelter problems of humanity. We are interested in organizing or participating in a resource cross-referencing system that deals with information on all facets of the shelter and furnishing problems. We are particularly interested in human methods of allocating services for basic utilities heating, water, electricity, and potential pollutants — and the most efficient and environmentally sound methods to produce same . . . . □

Bibliography on Yurts, Mongolian Life:
The Modern History of Mongolia, C. R. Bawden, N.Y., Praeger, 1968.
The Mongol Empire; Ghengis Khan, Peter Brent, London, Weidenfield & Nicholson, 1976.
A Lost Civilization — The Mongols Rediscovered, Walther Heissig, N.Y., Basic Books, 1966.
National Geographic articles on yurts: January, 1936; March, 1962; April, 1972.

Hoedad yurt frame

Geno

Sod Roof

by Robert Stowell

Approaching this house from the side, you see only a grass covered knoll on a high plain overlooking the distant Green Mountains. If the owners were very short of land for homesteading, they could readily graze a few sheep or goats on the roof.

The house is built on two levels, the lower level dug into the hillside, and the back wall is a rocky Vermont ledge. A pot belly stove warms the kitchen, and a long length of stove pipe saves a lot of heat that would normally be lost up the chimney.

Built largely of demolition materials from old barns, the house uses an unusual series of interior trusses spaced twelve feet apart. The size of the available beams has determined the overall dimensions of the house, about 28 feet wide and 32 feet long. It was built for just under two thousand dollars. Many of the timbers used have not been squared and were used as round or half round poles. The floor is made of twelve inch pine boards from an old barn.

Perhaps most controversial is the grass roof. The sod roof was placed over various layers of roofing and plastic on one inch sheathing. Some difficulty has been experienced in ice formation along the eaves during Vermont's extremely cold winters. The insulation value of six to eight inches (or more) of soil is not to be questioned, but adequate means of draining any surplus water is very important if you are going to use a grass roof. Weight must also be remembered. □

Vermont Cabin

by Robert Stowell

Henry Thoreau's dwelling was 10 by 15; this little cabin measures 12 by 16 and has housed a family of five through the summer months in Vermont.

Built largely of old barn timbers, the principal cost of four hundred dollars was for plywood roof sheathing and the asphalt shingle roof. The weathered grey vertical barn boards on the exterior seem to blend into the spruce and fir covered hillside.

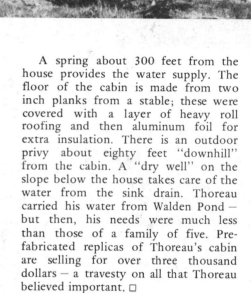

A spring about 300 feet from the house provides the water supply. The floor of the cabin is made from two inch planks from a stable; these were covered with a layer of heavy roll roofing and then aluminum foil for extra insulation. There is an outdoor privy about eighty feet "downhill" from the cabin. A "dry well" on the slope below the house takes care of the water from the sink drain. Thoreau carried his water from Walden Pond — but then, his needs were much less than those of a family of five. Prefabricated replicas of Thoreau's cabin are selling for over three thousand dollars — a travesty on all that Thoreau believed important. □

This mortise and tenon structure, built by Ian Ingersoll from salvaged barn materials, burned down in 1974. Winds blew down the flue when no one was at home and when Ian returned all that was left was "... smoking rubble, one foot tall...." He then spent two years planning before building again. Here is his account of rebuilding.

Starting Over

by Ian Ingersoll

After the fire when I began to build again, I decided to do things right for once. I dropped all ceilings to 7'6" around a single central wood fireplace, loaded the south wall with glass and, in all decisions concerning cost, I chose the cheaper (or "rawer") materials as the most pleasing all around. Local rough pines outside (with battens behind siding), local cherry on the floor, and rough-sawn timbers for the beam frame work. All left outside a year to weather grey. Foundation formed of rundown rubble on cascading ledge heading off in every which-a-direction,

my first head scratcher. Couldn't find any written material on forming footing on irregular ledge formations, so I figured out I could lay the beams I had sitting around into a log cabin-type affair outlining the foundation.

Next I faced the inside of this log cabin foundation form with plywood scraps and layed up my stone wall on the inside to my level chalk line. This gave me a level and square foundation (20' x 25') at a cost of $180 for sand and cement delivered, a good place from which to begin my house.

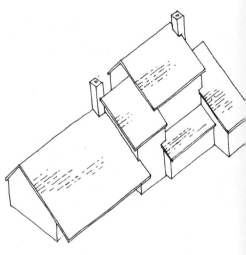

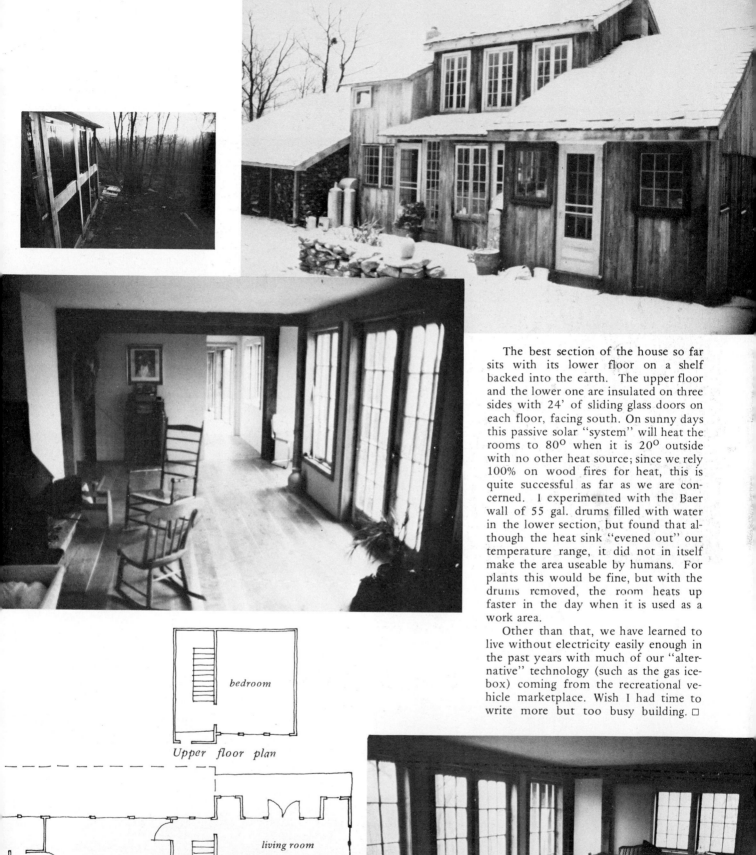

The best section of the house so far sits with its lower floor on a shelf backed into the earth. The upper floor and the lower one are insulated on three sides with 24' of sliding glass doors on each floor, facing south. On sunny days this passive solar "system" will heat the rooms to 80° when it is 20° outside with no other heat source; since we rely 100% on wood fires for heat, this is quite successful as far as we are concerned. I experimented with the Baer wall of 55 gal. drums filled with water in the lower section, but found that although the heat sink "evened out" our temperature range, it did not in itself make the area useable by humans. For plants this would be fine, but with the drums removed, the room heats up faster in the day when it is used as a work area.

Other than that, we have learned to live without electricity easily enough in the past years with much of our "alternative" technology (such as the gas icebox) coming from the recreational vehicle marketplace. Wish I had time to write more but too busy building. □

bedroom

Upper floor plan

shop

b

living room

dining

k

Main floor plan

164

Tent Top *by Ole Wik*

Distant views of our tent tops on barges at Goose Cove, Glacier Bay National Monument, Alaska.

A tent top is simply a wooden frame covered with a canvas wall tent. In the course of work in Alaskan national parks, my wife Manya and I have designed, built, furnished and lived in a number of them. In this article I will summarize what we have learned.

Suitability: Give a thought to a tent top if your shelter doesn't have to be readily moveable or completely permanent, if the climate permits tenting out, and if you don't have to worry about bandits or bears while you're away.

Materials: Wall tent, framing lumber, plywood, nails, screws, clear vinyl or mylar sheeting (for windows), hinges and latch (for door).

The Tent: Wall tents are all built on the same pattern, but they come in a variety of sizes, fabrics and colors, and are usually offered with either three or four-foot walls. We have found that a 10' x 12' tent, properly laid out, is entirely adequate for a couple. Depending on activities, finances, etc., two or more people may want a 12' x 14' or a 14' x 16' tent, and an individual might do all right in an 8' x 10'. In any case specify the four-foot wall, since even this will have to be extended. We pre-

fer a white tent, since the diffuse light is so pleasant, and we've had very good luck with 10 oz. double-filled fabric.

The Foundation: A tent frame is a charmingly light and airy structure, but it has to have a sturdy foundation. Make it plenty solid so that when you walk around the jars won't come crashing down off the shelves. I have built frames on pilings, using stout logs for beams, and also directly on the ground and on the deck of a barge. In all cases I used plywood for the floor because it is easy and fast to work with, easy to keep clean, and tight against mice and insects. A couple of coats of floor hardener ("gym seal"), available from petroleum bulk plants, will harden the surface against abrasion caused by sand that gets tracked in.

In dimensioning your foundation and frame, remember this very important point: tent sizes are given for the canvas before the seams are rolled. Therefore, if you make a 10' x 12' frame for a 10' x 12' tent, it will not fit at all. There doesn't seem to be any standard shrinkage factor, since different manufacturers roll their seams slightly differently, but 4" - 6" wouldn't be unusual for a 10' x 12' tent.

So lay your tent out on the ground, and measure the side and back walls carefully before you do any carpentry. Another pitfall to avoid: don't stretch the tent as you measure. If you apply tension to the canvas in one direction, it will gain extra stretch by shrinking in the direction at right angles, just as a rubber band gets thinner when you pull on it. I have fallen into this trap, and have had to knock a frame apart and cut it down a few inches because my tent, so carefully measured, would not fit when the canvas was stressed both horizontally and vertically at once. So shoot for a somewhat loose fit, remembering that the fabric will shrink a bit when it rains.

The Frame: Frame your walls 5'- 5½' high. (You will close the gap between the bottom of the tent and the floor later, with skirting.) Building the walls is easy: just lay out the sill, top plate and studs of one wall on the floor of the tent frame, nail them together, stand the completed wall up as a unit, and nail in place. When the walls are standing, square them up, add diagonal bracing at the lower corners, and nail them to each other.

Now draw the tent over what you've built so far and test the fit. If you did blow it, it is much easier to cut the frame down now, before you build the roof supports.

Wall tents always have a slit for a door right in the center of the front wall. But for a tent top, I prefer to frame the door way somewhat off-center as this gives added versatility in laying out the interior.

A good way to frame the door:

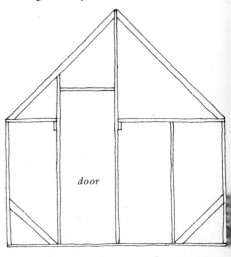

Front wall of tent frame

One stud, extending from the floor to the ridge pole, becomes one jamb; the door goes to the right or left of center.

Roof Framing: Manufacturers cut their canvas at a 45° angle and I make my rafters accordingly. For a 10' x 12' tent not to be exposed to severe snow load, I use two sets of interior rafters plus the two end sets, making four sets in all. I then add a long diagonal brace (say a 1" x 3" on both inner surfaces of the roof to give it the necessary rigidity and keep it squared up.

At this point, check the fit one last time. If everything is OK, set the tent aside and build in your furniture. It is usually easier to nail from the outside, before the skirting and tent are in place.

The Furnishings: A tent top lends itself very well to built-in (as opposed to moveable) furniture. Use your space as tightly as you would in designing the cabin of a ship. One paramount rule: always place the bed along the back wall. With the bed out of the way at the rear, everything else falls into place nicely along the walls, leaving a clear core area in the middle of the room.

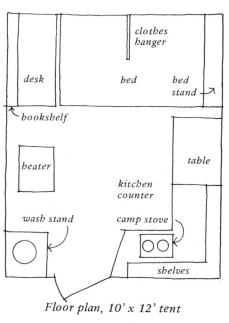

Floor plan, 10' x 12' tent

Notice that the bed doubles as a seat for the desk and table. Space under the bed is used for storage. We make drawers out of two kerosene boxes nailed together end to end and fitted with simple chair glides to protect the floor. We also build one or two lofts into the peak of the tent for things we need only occasionally. For hanging clothes, we suspend a 2' pole under the rear loft, over the bed.

A stool fits under the table. The kitchen counter is high enough for two-level storage: boxes on the floor and shelves above them for kitchen utensils. You will find yourself driving nails into the studs, top plate and rafters to hang all kinds of things.

The short part of the front wall, opposite the kitchen, is a good place for a wash stand, with the slop bucket underneath. (If you cut a circular hole in the surface of the wash stand to hold the basin, you can empty the water directly down into the bucket.) Finally, there is an open space along the remaining wall. We have always used it for a stove or space heater. *continued*

Tent top on a barge, Glacier Bay National Monument, Alaska. The fly of white canvas was recycled from another tent.

Kitchen in a tent top.

167

The Door: Always hinge the door to swing outward. Ordinary plywood makes a fine door. For a window, I simply cut a square panel out of the door, and then cover the opening with screening. Next I cut a smaller square out of the panel, thus making a frame which (of course) fits perfectly into the window opening. This frame, covered with transparent vinyl or Mylar, becomes a removable window.

Windows: Manya and I like a window right over the table. To make one, we cut the sides and bottom of the opening in the canvas, leaving the flap attached at the top. Manya then adds a strip of canvas to the flap, making it into a shutter which can be rolled down and tied in a storm, or else rolled up out of the way during fair weather. Into the opening in the canvas she sews a piece of screening. On the inside of the tent she adds a roll-down flap of transparent vinyl, and the window is done.

We lived in tent tops on barges while doing ranger work in Glacier Bay National Monument, Alaska.

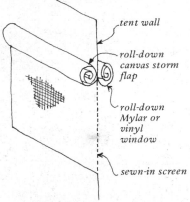

Window system for wall tent

Mounting the Tent: When your furnishings are in place, add the skirting (3/8" plywood is fine). Bring the skirting down below the level of the floor, so that rain will drain away, and make it high enough to allow the tent to overlap about 3". Then pull the tent over the frame, starting from the rear. Pull it down snug, and secure it to the frame with plenty of staples along the lower edge. (A batten along the lower edge will keep drafts and mosquitoes out.)

When you get to the doorway, you will have to cut the canvas and fit it to the frame. Leave enough material so you can draw the fabric around to the inner surfaces of the jambs, and turn a hem under before stapling it down.

Waterproofing: A new tent should be adequately watertight as far as the walls are concerned, but the roof may admit droplets in a driving rain. One way around this is to mount a fly.

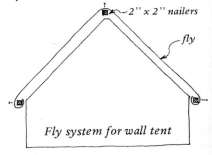

Fly system for wall tent

Our most successful fly was made out of the roof of an old tent. It was white, and gave us rain protection without cutting down too much on the light that came through the roof. In a wind-free area you may be able to get by with transparent polyethylene sheeting; otherwise you may have to use a regular tarp, which will darken your dwelling somewhat.

The other approach to waterproofing the roof is to use a commercial preparation. We have used Thompson's Water Seal with very good results. We mounted the tent first, and then applied the sealer, full strength, with a paint sprayer. Treated fabric doesn't "breathe" the way it does when new, so spray only the roof, not the walls.

Penny Rucks at the table in the tent top.

168

At Right: *Keith Jones's tent top cache, under construction. Keith and other fishermen often carry very light tent frames, made of spruce poles, in their boats when they go to their camps at the beginning of the salmon season.*
Below: *Doing string tricks in the tent top.*

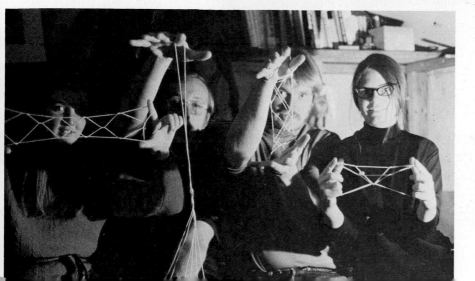

Decamping: At the end of the season, you remove the staples and the battens, then roll up your home for the winter. Next spring you just sweep the dust, leaves and alder cones out of the frame, put the tent back on, wash off the oil-cloth on the table, bring in your personal gear, and you're in business.

When you wake up the next morning, you listen to the dawn chorus of the birds as you watch the shadows of the leaves dancing on your roof. From the shadows cast on the wall by a particular sapling, you have an idea of what time it is. The breeze through the screens brings you morning air that is fresh, dewy and full of promise. □

Back to the City

In 1959 George Abernathy retired from the Navy and he and his wife Nell moved to 40 acres of forested land in the mountains near Oroville, California to build a home. At an age when most people are thinking of retiring, George Abernathy spent seven years building an 1800 square foot house, using stone from the site and Alaskan-milled pine beams from nearby woods. Now, 20 years later, they have their soundly-constructed house up for sale and are planning to move back into the city, where they will be closer to other people. "... sanctuaries are one thing, but complete satisfaction for all your needs may not be provided in a sanctuary. You can only take so big a dose of contemplative life."

George and Nell Abernathy

L.K.: *How did you go about deciding what kind of house to build and where to get your information about it ... you didn't consult an architect?*

George Abernathy: No, the whole thing is a one person operation and I consulted, really, no one except for the data I got from the Government Printing Office and other information such as that.

Is the site away from town?

Oh, yes.

When you moved up there did you want to get away from urban or suburban areas, was that the idea?

The idea first, before the place took us over rather than we taking it over, the ideal was a place in the mountains accessible to the sea and to the snow. Well—this is a little far from the sea, but it's close to the snow and it is in the mountains. And it had to be a place with an atmosphere of sanctuary where you could get away from all the noises. The noises we hear here are the airplanes that go over, and occasional vehicles on a nearby county road. And it's a place where you can relax to the extent of taking your clothes off and walking around bare-ass in bright sunlight if you want to. And it's a place to which friends are attracted to foregather when they have time to be relaxed and at ease.... If a person gets here and after a while can't realize that it is a good place to cast their worries aside temporarily, why, they're not very sensitive.

You've been up there about 20 years?

Yeah.

And now you're considering moving back to a more populated area ... do you feel like being around more people?

Well, there are several factors entering into it. The principal and the practical one is that we're both, my wife has just turned 80, and I'm almost that and we're still in good health and we could stay on, but as time goes by it is not the place for two old people to be alone in in times of rough weather ... We can still manage snow shoes and skis, but there are occasions, a few years apart, when we get heavy snows here—one time we were shut down 10 days. No sweat, we were prepared for it—all the emergency necessaries, emergency motor generator, plenty of wood, plenty of food ... But that is the principal reason, that age creepeth upon us, and then too—a social comment, I hate to make it—but there are only the two of us and we've told one another what each one knows and believes several times over again; we have no intellectual companionship.

Well, it interests me because so many people are painting an idyllic picture of moving to the country and getting away from it all and often the people who write about those things are either making a living by what they're writing, or they're writing their article on a morning when they've just had a cup of coffee and the sun's just come up. That aspect of your experience interested me, these things you just mentioned.

Let's face it, my friend, there's no simple answers to anything excepting the most profound questions; that is, just ordinary questions which you seem to have a simple answer to, view them with scepticism.... We're all caught up in the solipsistic dilemma: there are as many universes as there are individual intellects trying to describe them.... You can only look at things from your own view point, however charitable ... Well, when you're looking at it from the other fellow's viewpoint, you're estimating what his viewpoint is before you look at it *from* that, so to speak. With my studying with Dr. (Gregory) Bateson and thinking about his communication theory, a person in isolation, *especially* one who has developed any sort of intellectual complexity needs nexuses outward always, through which he gets nourishment from others of like ambience, so to speak. A person is a constellation of communication relationships and if you don't get nourishment through those channels of communication now and then you starve to death.

Had you thought about these sorts of things for quite a while and discussed them before you decided to move back into the Bay Area?

No, those things are cropping up vaguely, and Nell's telling me she's running out of things to do, and I'm just finishing off this place and she's saying have you considered what you're going to do when we go back to the city, you'll be climbing the walls for something to do? Well, I've got several projects as far as reading and writing are concerned that I've wanted to do and I've tentatively thought maybe I might go back and be a relief teacher at the San Francisco Institute of Fine Arts, or perhaps get interested in city politics.... sanctuaries are one thing, but complete satisfaction for all your needs may not be provided in a sanctuary. You can only take so big a dose of contemplative life.

Have you run across any other young people who moved out there in recent years ... A lot of people fantasize about what it will be like to get away on their 40 or 160 acres and then they get out there, away from other people for the first time ...

There have been two or three couples we have observed and they've gone now, but they were of the kind that, "Boy, I am going to get out of this awful meat grinder of civilization and get back to the simple, sweet, direct life." They got old houses and "We're going to do this over and we're going to upgrade the facilities ..." and they worked their butts off for a couple of years. And water pipes would freeze and the houses would be cold in the winter, the places would leak, and the animals wouldn't behave properly, and they'd get a toothache and they couldn't hack it. You can ask nature for only so much and she demands so much from you and they didn't know how or couldn't allocate sufficient time to live under these so-called primitive circumstances.

As far as your house goes, it's a stone house with Alaskan-milled timbers?

Well, the basic idea was to make it simple and good at the outset, using all the best of the old and the new. Well, the oldest houses, of course, were caves and stone and mud, and so forth. There had been a mine on the place with copious tailings out of which I worked sufficient loose stone of random, rubble type, as they say in the masonry manual, which means irregular shapes. There's no drift or square stone in the walls at all and I found out, to my dismay, that it takes about ten times as long to build that way. At the outset I was trying to recover the sand for the mortar from the mine tailings, which was beautiful stuff fine white quartz, but after a little while I discovered that it was taking too long to clean it up. It had to be hauled in to the site, screened and washed to get the clay out of it. So I broke down and bought sand, and the footings of the house are poured concrete and the walls and the footings are one foot thick and taper to 8 in. at the rafter seats. It's a one story and the roof is supported on 8 x 8 handmade beams, 44" centers with cardecking-strength roof sheathing, and V-rustic cedar ceilings with 2" insulation between the roof and the face ceiling. Roof is Johns Manville cement asbestos shingles, which are fireproof, and the inside of the house is 24 ft. x 35 ft., which is just short of the golden mean proportion.

What was the reason for the golden mean? You knew of it and thought it would make a nice proportion?

Well, throughout time it seems to have been a *comfortable* one. The square is monotonous, and long is tenuous. If you work these things around, well, to the eye and to the inner senses, this is just a good shape. And whether that was tendentious on my part or not, I don't know, but it seemed to have worked out that way. Then the height of the windows . . . I'd got the floor laid and before I started setting the forms against which I laid the masonry, I had to place the boards for the windows. So I put a chair alongside where the wall was going up and had my wife sit down in it and then made a light frame and held it up or down and let my wife look out through it until she got the best picture. And then we set the windows at that height — to suit the convenience of her view.

" . . . we spent our first night in a Norwegian igloo tent and it rained. And the next morning I got up and was getting something hot to drink and Nell said, 'Darling, do you know what we've done?' And I said, 'No, what have we done?' She said, 'You've just bought yourself 20 years hard labor.' She was right."

Did you start out to build the house using native materials?

Yes, that was the whole idea.

If you were going to start over again, do you think you'd change anything? Do you think that's still a good approach?

Well, it's time consuming. If I started out again, knowing the time I had, I'd probably do it another way and buy a little assistance to save time. First I'd excavate a nice basement that would be insulated and then off from that, accessible to it through a door, a storage cave for preservable vegetables like squashes, potatoes and apples. With that as a footing, I'd make a balloon-type framed dwelling using good insect-repellent materials, say redwood sills and 2 x 4's and finish it on the inside with standard gypsum board, and on the outside it would be a choice between board & batten cedar to let it take the weather, or embossed aluminum siding that would work in with the trees and the land. And then a choice between an embossed aluminum roof with a high pitch or stay with the Johns Manville cement asbestos shingles for fire protection reasons.

In other words you'd build a wood frame house?

Well, it would save time and it would also give better insulation. Stone has a

great deal of thermal inertia, it takes a long time to get hot and a long time to get cool and if you build it for energy conservation, the aluminum type framed house with good heavy insulation would be far better, because if you use 2 x 6 studs instead of 2 x 4 and put a net five inch insulation around the house and same amount on the roof you would get something like a refrigerator: it would be hard to get it hot in the summertime or cold in the wintertime.

Well, that's what they've been doing back East lately. They use 2 x 6 studs, 24 inches apart and put 6" insulation in between.

I've got a neighbor who built a house somewhat on that order, and not only is it heat-inert, but it's extremely silent. His nearest neighbor came up one day and in the middle of a rainy windstorm and they came into the house and sat down and the neighbor said, "My God, it's stopped raining." They got up and went out and the storm was still going on. But the noise associated with the storm was absolutely muted by the house.

Well, that's very interesting — what you did, and when you began it.

It seems to us a very commonsense, unexciting way to manage one's left-over time.

You were almost 60 years old when you started?

Yeah, that was about it. I'm seventy-five now . . . .

Not very many people start building a house at that age. Do you have any secrets as to your good health?

No, I suppose I'm pig-headed and opinionated and contentious. Just between me and thee. When I went on the retired list I was full of anger and frustration at the way things were going and I'd heard somewhere that the best palliative for that sort of thing was to do a hard piece of work. When Nell and I bought this place, she recalls, she reminds me now, that we spent our first night in a Norwegian igloo tent and it rained. And the next morning I got up and was getting something hot to drink and she said, "Darling, do you know what we've done?" And I said, "No, what have we done?" She said, "You've just bought yourself 20 years hard labor." She was right.

Well, it sounds like it has worked out all right. Especially since you have something which is of value now, you can pass that along to somebody else or almost trade it in for a place closer in to the city.

Right. Well, one thing that can be said simply is that our 20 years effort hasn't damaged nature any, and could have helped it. □

Interiors

Cities

In American cities, people are finding ways to stop the bulldozers, resurrect abandoned buildings, repair and maintain older housing, and create their own living space.

Derelict structures, formerly fire hazards or wino crash pads, are being cleared of rubble, rotting timbers and broken fixtures and are being refurbished as healthy, well-tended living quarters. Homeowners in some cities are being given aid to maintain and repair existing housing. Neighborhoods once plagued by dangerous, ugly, abandoned buildings are springing back to life with rehabilitation underway.

Some of these urban projects could well be forerunners for similar programs in other cities: revitalizing depressed areas and providing jobs, income and housing for those who need it most. Common to these projects is the concept of people working together to provide their own shelter, rather than hiring others. Not only is it cheaper, but by cleaning up, repairing and rehabilitating, they are learning valuable skills, taking a vital interest in helping themselves and others, and are later able to cope with emergency repairs and building maintenance. Funding for the programs has been provided in various ways: by private banks, the federal government, or the city.

In Cambridge, Massachusetts, two rehabilitation programs are taking place: one, aimed at improving existing homes, offers homeowners small grants and low-cost loans to paint, repair, and maintain their own homes. The other, called the Work Equity Program, makes badly deteriorated buildings available to low or medium income families; the families are allowed to buy these buildings by contributing their own labor (a year or 18 months work) instead of a cash down payment.

In the South Bronx, a well-organized, hard working group called Peoples' Development Corporation (PDC) have cleaned out and rebuilt an abandoned six-story brick building built in the early 1900's into a 28 unit apartment building in which PDC members now live. One of several active urban homesteading groups in New York City, PDC has started an ambitious program to stabilize first one block, then nine blocks, with the far-reaching ideal of rebuilding a 40-block section in one of the worst neighborhoods in the nation into an "urban village" of 10,000 people.

The first phase of their program is impressive: members have earned a share in the building by contributing 500 hours "sweat equity" labor in lieu of a down payment; they have learned construction skills while providing their own housing; their work has revitalized the neighborhood; solar heated water, recycling and a food cooperative are realities; and day care, health care, midwifery, nutritional consultation and a small park and playground are either underway or now functioning.

Several things are apparent from studying the programs in Cambridge and the South Bronx: low income people who are unemployed and have inadequate housing can work to clean up and rehabilitate abandoned buildings and provide new and decent housing. While doing so, they can learn valuable trade skills, occupy otherwise unproductive time, often (via grants) earn wages, and have the long-range benefits, experience and pride in doing something for themselves. The cities benefit by having decent housing created out of the rubble, and the buildings back on the tax rolls. The neighborhoods benefit by an infusion of vital energy in formerly depressed areas. And the participants benefit by discovering inherent, yet latent capabilities.

Perhaps the most important feature of these programs is that they started in the neighborhoods; they were not the result of city or government initiative (although the city or government has cooperated) or political or administrative planning. They started with individuals getting together, discovering and reinforcing common goals, organizing, finding the financial backing (private or public) and following up on the initial concepts with dedication, dependability, political awareness and hard work. □

175

What *does* work is now being demonstrated in the city of Cambridge, Massachusetts, in several housing rehabilitation programs that have been underway for eight years. A non-profit corporation, Homeowner's Rehab, and the City of Cambridge cooperate in running several programs to assist low and moderate income families or individuals to either purchase and rehabilitate run-down buildings, or repair and maintain existing homes. In the following discussion, Mel Gadd of Homeowner's Rehab talks about housing rehabilitation and neighborhood improvement in Cambridge.

L.K.: *How big a city is Cambridge?*

Mel Gadd: It's a population of 100,000.
What is Homeowner's Rehab?

The corporation runs two programs. One is a home improvement program, which is directed at working class families who already own their property. Pushing a self-help approach where we're willing to teach people how to do anything themselves, because that's one way they'll save money. We come up with some financial assistance in the form of a grant where we'll pay for part of the work or subsidize a conventional home improvement loan, plus give them a lot of technical assistance related to fixing the house up; property management, financial counseling, things like that.

The other is the Work Equity Ownership Program, which is directed at getting families who normally can't get into ownership situations, into ownership, the biggest problem being that they can't come up with that 20 to 30 percent down payment that banks require. And what we do is we substitute their doing labor on the house for that down payment. Plus, at the same time, we're taking over really burnt down, abandoned, ripped apart absentee properties and putting them back together, getting owner-occupants into them, and getting them back on the tax rolls.

Do they live in the house and pay rent for a period of time?

Right. Once we buy the house and we do the major work, major structural systems, mechanical systems, the family

starts to do their share of the work that's agreed upon at the front end, and they move in and get a year or year and a half to do that work.

And they pay you rent while they're doing that?

Right. They pay us rent to help cover our carrying costs on the building because we borrow construction money to buy and rehab the building. We're paying the taxes and the insurance.... Once they complete the work then we get them a permanent mortgage which, in effect, is at 100% financing. They don't come up with any cash at all. That's the commitment the banks have made to the program.

To bring it up to date, roughly how many families have been involved and how many houses or buildings?

Approximately 50 units under the Work Equity Project and we're way over 1000 units in the Home Improvement Program.

What income level people are you working with?

The Home Improvement Program works with anybody up to income grouping of around 15 or 16 thousand, and that's for a large family, about 6 to 7 people. The Work Equity Program deals with families from roughly 10 to 15 thousand and two people working. The only way we're servicing other low-income families under the ownership project is like in our condominium which we did a couple of years ago, where we were able to bring the unit prices down far enough

to get a single person with an income of $6,000 in.

People are buying those condominiums for how much?

Well, this one 8-unit Condo we did, the average price was $10,000.

That sounds fairly miraculous.

Yeah, especially since they've been appraised at 20-something.

Are you getting any heat from builders or unions?

No. The unions don't bother us because they just haven't been able to get into the rehab game. And they know no one would do it if the unions got involved; the cost would be too high. No one has complained about the program; the city loves it because we're taking burnt-out buildings that have not been paying taxes and putting them back on the tax rolls. We don't ask for any special tax benefits ....

You're not competing with any vested interests?

No. Although it's getting a little harder for us to buy property because the market here is going up so quickly and we're starting to not be able to afford houses because young professionals coming in are willing to pay really heavy prices.

Maybe you'll bring about your own demise.

Well, it's a potential. What we've considered now as another alternative is to do a couple more condos, because we can keep the unit prices down by doing so many units in a building.

continued

Parts of a House to Check for Defects
1. chimney
2. dormer, and its connection with house
3. roof shingles
4. gutter, and its connection with house
5. connection of gutter with downspout
6. porch, particularly any sag, bulge, or lean; the connection with the house and the condition of the roofing and gutter
7. window frames, especially at corners where pieces are joined
8. window glass
9. walls, particularly any sag, bulge or lean
10. walls, shingle or clapboard
11. foundation
12. fence

Have neighborhood blocks really improved through this program? Can you drive around and see improvements?

Right now we're doing a big 12-unit one that's having a major impact on a street. It's a fairly small block and this was *the* major parcel on the block that caused it to start going down and now ... the neighborhood is starting to go up. A lot of the Work Equity buildings we've done have been scattered around the city where they're the worst building on the block. A number of them have been the only burnt-out structure on a basically sound working class street that affected everybody else and now that we've redone them, everybody's thrilled.

How did the Homeowner's Rehab get started; did individuals start it?

It was neighborhood board members from the Cambridge Model Cities Program along with some people from the Cambridge Corporation, a non-profit corporation.

So the neighborhood people presented their idea to the city? How would an individual start something like this in any other city?

Get a block organization or a neighborhood organization.

Start in the neighborhood?

Yeah, that's the best way. Get a good neighborhood group formed that's really interested in doing it and incorporate and then go look for the money. Identify some bankers in the city that you can

get behind you because the program doesn't work without banks and then start looking for the administrative money to pay some staff. But if you've got a strong neighborhood group that's willing to work and a couple of bankers who are willing to make a commitment, the rest of it is going to fall into line.

Anything you can think of about this program that would be different in other cities?

We feel this is a good model. But, you can't just take our set-up and do it, say, in the Bronx. You've got to say "Here's the model, what are our specific problems, what should we change to make it fit us perfectly?" You can't just take the glove and put it on someone else's hand, you've got to mold it a little bit. Other places have other problems. We don't have some of the problems that other cities have ....

Anything else to add to bring us up to date?

... We are busier now than we have ever been. The goals of the program are getting there. But ... self-help is the thing we push. Families see that they're getting to own a house without coming up with any cash. O.K., we feel that we're tying them up for a minimum year to 18 months where they're learning to do a lot of repairs and improvements on their own property.

During that period you're seeing if they can work ...

Right. A lot of the other government-

ownership projects have just given people houses and walked away from them. And the first time a major problem comes up, they don't know what to do and they lose the house. Statistically you'll find that there are a lot of HUD foreclosures on those ownership projects; we feel that the counseling part of the program, where they're learning how to do the work, prepares them for ownership better in the long run. The complete program: that's the important part. People may think they're getting something for nothing, but sub-consciously they're being bombarded with a lot of information and are being prepared to deal with the house in the long run, so that if, in the future, there's a leak, they either know who to call or how to fix it themselves. Plus people are less likely to rip apart a house where they've done all the work and we've found they end up being a lot more picky about the quality than we are because they're doing it.

That's great. You've got people doing something quite natural.

Sure ... owner-built homes and owner-repaired homes are nothing new, it's just that no one has ever approached ownership programs for working class families on that same basis. No, they're saying, "You've got to give it to them, period, and that's all you've got to do," and that's really a mistake.

Well, you make them an owner-builder instead of an owner, and they'll love their home.

Right, in the long run it's far better. □

Funding

The Work Equity Program is presently being financed by both public and private funds. $35,000 of FY1976 Community Development Block Grant monies cover the salaries of Homeowner's Rehab staff. $25,000 of additional CD monies allocated to the City of Cambridge finance a Work Equity Exterior Demonstration Project as part of a Rehab Demonstration project. These supplemental funds frequently finance siding, shingling and painting work and reduce the cost of the permanent mortgage obtained through the Work Equity Program.

A number of local commercial banks are supporting the program by making short term construction loans available to Homeowner's Rehab, Inc. for the purchase of deteriorating structures. Homeowner's Rehab pays interest on the 18 month loan; no principal is required. Families are able to secure permanent 100% mortgages from savings and commercial banks for the purchase of the structure from Homeowner's Rehab after completion of their repairs. The terms of the mortgage usually include an average interest rate of 9%, to be paid over a period of 25 years.

Since the program began in 1973, Homeowner's Rehab has been able to secure an average purchase price/unit of $6,802. The average permanent mortgage/unit obtained by families has been $11,249. The average value of work equity performed by participating families has been $1,053. The average rent paid by the family to Homeowner's Rehab during their period of work equity has been approximately $160/unit plus utilities.

Performance

Structures have been between 70 and 100 years old. Eight of the 17 structures purchased by Homeowner's Rehab since 1973 have been absentee owned. The remaining structures were owned primarily by elderly people who had lived in the building for many years, but because of deteriorating energy and funds found it impossible to maintain the structure.

An 8-unit condominium was developed through the Work Equity Program. The Broadway Building Condominium is the first conventionally financed condominium for low and moderate income families in Massachusetts. Seven of the 8 units have been sold to Cambridge residents. The condominium units were sold for $8,000 - 12,000/unit (2-3 BR). Owners of the units are managing the condominium themselves.

Bootstrap

Families of low and moderate income are able to secure reasonable permanent mortgage financing not ordinarily available from conventional sources. Appliances and some materials are purchased wholesale at a considerable discount (30 - 50%) which reduces the cost of the permanent mortgage. Some materials used by Homeowner's Rehab in the major rehabilitation work are also purchased at discount. A member of the Homeowner's Rehab staff acts as general contractor, thus reducing the cost of rehabilitation. The value of the work equity usually equals 10% of the actual purchase price of the building, or considerably less than the down payment currently required on a conventional mortgage.

Included in the rehabilitation costs of the permanent mortgage is approximately $1500 designated as the Family Budget. This sum of money is used by the family to purchase cabinets, stove, refrigerator, lighting fixtures, paint, tiles, wallpaper, and materials used by the family in performing their work equity.

In practice, many families participating in the program have lived in Cambridge for most of their lives. They are tenants who have never owned their own home and are primarily middle-aged couples with a number of children. The incomes of participating families seem to fall between $10,000 - 15,000/year. □

Homesteaders in the Bronx

by Ned Cherry
Photographs by Mark Haven

The People's Development Corporation is a group working in the South Bronx, New York City, to rehabilitate buildings such as these. Having transformed an abandoned, trash-filled six-story brick building into a clean, well-planned 28-unit apartment house, PDC members have now started rehabilitation of five more buildings.

Housing rehabilitation in New York City began in the late 1960's under the municipal loan program, whereby a building owner could obtain a low-interest, long-term loan from the city, make major or minor repairs to a building, increase rents and generally not involve the tenants living on the property. This program ended due primarily to fraud, scandal and general profiteering, and the city learned that to improve run-down buildings and neighborhoods required greater involvement from the people inhabiting them. Thus when the program re-emerged in the early 70's it included such innovations as *cooperative conversion* and *sweat equity*. Cooperative conversion enabled tenants to actually own a part of the building after rehabilitation, and sweat equity allowed tenants to substitute their own labor in lieu of the standard cash down payment required on long-term rehab financing.

The need for rehabilitation emerged primarily as a result of two factors. One, the mass abandonment of apartment buildings after owners failed to pay

182

Ned Cherry

taxes on them, leaving the city with close to 20,000 buildings in tax arrears, or *in rem*. The second factor was a combination of the high cost of new housing construction and the lack of government commitment to subsidize this housing for low and moderate income families. Generally speaking, rehabilitation costs between $10-20/sq. ft. whereas new construction was up to $35-45/sq. ft. The city also realized that to retain some character and scale in neighborhoods, it could not continue to bulldoze old buildings and substitute rows of inhuman building blocks.

In response to the new housing rehabilitation program, many community-based organizations developed with the express intent of obtaining buildings from the city, forming cooperatives from remaining tenants and other neighborhood residents, and rehabilitating these buildings to suit the needs of the new building owners — the tenants themselves. Among the first of these organizations to form in New York were Adopt-A-Building on the Lower East Side of Manhattan; U-Hab, under the sponsorship of St. John the Divine Church in Morningside Heights; the Renegades, a former East Harlem Street gang; Los Sures, an Hispanic-based community development organization in the Williamsburg section of Brooklyn; Housing Conservation Coordinators in the Clinton (formerly Hell's Kitchen) area of West Side mid-Manhattan: and the Peoples Development Corporation (PDC) in the Morrisania section of the South Bronx.

PDC was formed in late 1974 with the assistance of U-Hab and under the leadership of Ramon Rueda. Rueda and others from the neighborhood selected an abandoned 6-story apartment house on Washington Avenue, with the intent of stabilizing one end of a 3 or 4 block long area of the street, there having been a relatively new housing development up the street stabilizing the other end. Once decent housing was provided at both ends, they planned to work on improving the buildings in between. *continued*

Ron Diamond

PDC requires 100 hours of "sweat" to assure seriousness of purpose and 500 hours as down payment on an apartment in the building. They also require a commitment to the neighborhood in that anyone working for PDC must live in the area within a year's time.

Most rehab work in the Bronx is gut rehabilitation, meaning that once the building is cleared of debris and rot, there is little left save outside masonry, interior bearing walls, maybe some stairs, and a portion of the wood floor joists that have survived roof leaks, fires, faulty plumbing or infestation. Windows and frames are beyond repair, doors demolished or stolen, gas, steam and water piping long gone (stolen), bath and kitchen fixtures broken, and heating and hot water systems broken or missing. What makes the prospect of turning such a nightmare into living space even thinkable is the high cost of new construction in recent years, and the fact that many of the abandoned buildings are of sturdy construction—better than anything that could be built new.

Their first task was to clear out the years of accumulated debris in the building, ranging from old refrigerators to dead dogs — a six month effort. With the aid of a seed money grant of $19,000 from the Consumer-Farmers Foundation, they were able to obtain technical assistance in developing plans to present to the city agency.

By September of 1975, NYC's impending fiscal crisis was evident to the Housing Administration, which tried to back out of its prior commitment to the PDC project. Rueda and 30 other PDC members threatened to sue the city for breach of contract, occupied the chief administrator's office, were jailed overnight, but when the case came up before a judge it was dismissed and the city finally sold PDC the building for $1. Along with control came a $311,000, 30-year municipal loan for rehabilitation plus a $220,000 job training grant

Ned Cherry

Ron Diamond

"This is our baby, this bond keeps us together."

Ruben Rivera

"I was just hanging out, drinking, on welfare, doing nothing. Then I started rapping with Ruben. He turned me on to this, said it might lead to a job. He got me interested in the program. So I started putting in my sweat. Man, I stopped counting, I put in so many hours. In five years this neighborhood is coming up. Now it's joyful when I'm working here. I brought more people in."

Edward "Moose" Holmes

from the city's criminal justice coordinating council program. This program enabled PDC to train and pay some of the neighborhood people who had had previous brushes with the law.

Although PDC had the manpower to attack the rehab project, they lacked the necessary construction skills. Thus when they took on skilled carpenters, electricians and plumbers, it was understood that their members would be able to apprentice under these technicians in order to learn their trades. Thus, the 18 month construction period resulted in not only providing PDC with new housing quarters involving cooperative ownership, but with its members learning the skills so they could rehab more buildings in the neighborhood. As the first rehab project was proceeding, PDC was becoming a more sophisticated organization and developing broader goals for itself. *continued*

185

Recycling bins in the hallway of each floor. Six or seven years ago when people would discuss recycling, use of plastic containers, etc. in NYC, some one would always say, "That's great, but try to tell it to the people in the South Bronx." Yet that's exactly where it's now happening, only with a lot more sense and dedication than on the Upper West Side.

One of the four solar collectors on the roof of 1186 Washington. The system provides approximately 45% of the building's domestic hot water during the winter and about 70% in summer; this allows them to shut off the boiler in summer. For 28 apartments this constitutes a considerable energy savings.

"I don't know if the PDC movement is democracy, communism or socialism, but it works."

Paul Reyes

Its first phase — the 1186 Washington Avenue project — has progressed to a second phase, five building rehabilitation plan. The original concept of a nine block stabilization has developed into a 40 block master plan including recreational facilities in vacant lots, nutritional programs in conjunction with local health clinics, plans for 24-hour day-care centers, building management programs, etc. They have developed their own planning and design unit which works with prospective tenants to assist in the layout of their own living units. They have set up a cabinet shop in a nearby abandoned brewery storage warehouse to produce kitchen cabinets for the rehab projects, as well as teach cabinet-making skills to interested neighborhood partici-

pants. All these programs are striving toward broader community participation without which one or even six rehabilitated apartment buildings will not turn a declining neighborhood around.

Herein, I believe, lies the key to PDC's success thus far. They have emphasized community participation with an almost "something for everyone" philosophy, gained the recognition and respect of the neighborhood residents and thereby are somewhat ensured of at least moderate success. By offering people in the area opportunities for housing ownership, development of useful skills and a chance to affect their own environment, PDC has created the necessary ingredients for a workable self-sustaining community. They feel that a "village

Ned Cherry

Above: *Peoples' Development Corporation's rehabilitated 28-unit apartment building in Morrisania, South Bronx; note solar water heating panels on roof.* Below: *PDC members on roof of their building.*

structure" of 10,000 inhabitants can be achieved in the Morrisania section of the South Bronx. Toward this end, they have incorporated solar heating in their first rehab project; have developed and encouraged recycling centers in the building and neighborhood; are experimenting with raising fish for food; planning a solar heated greenhouse in the area; setting up an earthworm raising center for eventual sale to the Bronx Botanical Garden; establishing a recycled building components storage facility; developing programs in building management and maintenance, nutritional care and midwifery, and a range of activities aimed at involving as many area residents as possible. □

Amsterdam House Boats

by Paul De Leenheer
Photos by Hugo Schuit

In Holland's Golden Age, the canals of Amsterdam were busy, full of life, and served as the home for those who worked on the waterways. But with the industrial revolution, sea cargo went to the larger port of Rotterdam, and the canals of Amsterdam were no longer the vital arteries of Holland's transportation system.

Now, there are about 10,000 people living on the water in Amsterdam, on about 2300 boats. On the Amstel River, where once there was one row of boats along each *quai*, the boats are now two or three rows thick.

There are two categories of houseboats according to the law, the legal and the illegal. The legal ones have a license that allows them to hook up to the city's water and electricity systems. They are usually pre-fab houses on concrete barges, anchored together in neat rows. There are about 1000 of these, and the rest are the so-called illegal boats, making up the most colorful and lively aspects of this watercity.

The city closed the canals to new houseboats in 1975 and many of the illegal houseboat dwellers feel threatened. Very few boatpeople want to leave their boats for an apartment, even though the city offers to buy their boats and pay for them to move to a flat. In fact, some of the owners have sold their boats to the city, then used the money to buy better boats. □

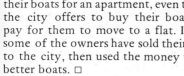

Norfolk Street

Millions of low-income elderly Americans are today faced with a grim choice of housing prospects: a) public housing projects (usually drab and uniform) for those who can manage their own households, finances, cooking and cleaning; b) nursing homes for those who cannot function independently and need care and supervision, and c) urban residential (skid row) hotels.

Yet many older people do not fit into the categories defined by these housing types. Many are single, cannot keep up their homes any longer, are on a meager income yet are not so incapacitated as to require a nursing home, nor inclined to the loneliness and frustration of either a public housing project or a skid row hotel.

A project in Cambridge, Massachusetts, conceived by architect David Conover and built in 1975, provides an alternative to these choices. In a former convent converted to what is being called "congregate housing," residents share kitchen, eating, social rooms, even bathrooms rather than live in separate isolated mini-apartments of their own.

"Even though there are a few hitches" says Conover, "the people seem to be happy, and it really is trying to accomodate a population that shouldn't be by itself because of a certain built-in isolation, or loneliness, or physiological problems, to some degree as an outgrowth of a sense of uselessness . . ."

At 116 Norfolk Street there are 39 units, averaging 250 square feet each. There are six to twelve apartments on each of three floors. Each has a bedroom, usually a sitting room, a toilet and a wash basin, but the bathroom is down the hall — one for every three apartments.

" . . . a lot of older people have arthritis . . . and often need some form of assistance in their bathing. The shared bathing facilities make it much easier for someone to ask for help. If they're intending to take a shower or bath and feel they need some help, it's much easier to invite somebody into a room that's being commonly used rather than into your own unit . . ."

Outdoors, a large walled garden, once the cloister of the convent, has space for residents to plant their own gardens. In-

Central to the congregate housing concept are common dining and kitchen facilities. Even though it is not true that illness necessarily accompanies the process of aging, illness does accompany malnutrition. Our cities are filled with elderly poor who do not eat properly. Some, with their meager income, prefer to drink an evening away than even contemplate eating alone. Many diets consist almost wholly of ready-prepared carbohydrates such as doughnuts, cupcakes, cookies and crackers; milk is markedly absent from most menus; meats and vegetables are present in minute quantities.

Many people living alone simply do not want to prepare their own meals. Men, in particular, tend not to cook,

and will eat out whenever their budget allows. It has been estimated in one of the newer housing projects for the elderly that upwards of 25 percent of the residents have not used their own kitchens; snacks have replaced meals; eating from cans is commonplace.

This phenomenon, malnutrition — self-imposed or, at least, unwitting — is almost wholly an outgrowth of social isolation and loneliness. Eating is a social act; even cooking makes sense (and is done with pleasure) if it is done for more than one and is shared. Yet, with our current belief in self-containment, the typical isolating apartment denies many single individuals the impetus for maintaining personal nutritional care.

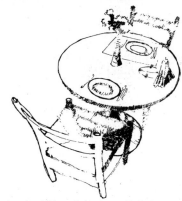

doors, the building has been left intact wherever possible, with even an old bread warmer from the convent kitchen still in place. Fine old wood floors were refinished, not covered up, and the lanterns from the former chapel now hang in the dining hall. Doors and door frames from the nuns' cells were removed and used again, to keep the kind of traditional detailing older people can enjoy.

The renovation was undertaken as a private project, then transferred to the Cambridge Housing Authority. Rents average $75 per month.

One hot meal (for $.50) is served in the main dining room each day to those who wish it; other meals are prepared by tenants in the common kitchen/dining areas on each floor. The important thing is that residents have the choice of eating out, eating with others in the main dining room, or fixing their own meals.

Part-time jobs are available through a "home-industry" program. Jobs have been contracted with local industry for half a day, five days per week. This work is often cyclical and coincides with the peaks and valleys of a company's business flow. This works well for elderly people, who are flexible in adjusting to this alternating work load. The jobs not only provide income, but help the elderly maintain a sense of usefulness, dignity, and self-worth. Architect Conover feels that it is important to locate a congregate housing facility for the elderly in an area which is not isolated, but within easy walking distance of stores and public transportation.

Social researcher Sandra Howell of Cambridge estimates that perhaps 15% of all older persons might prefer a congregate arrangement like this to living by themselves. She predicts congregate housing will best serve two types of people: those who are somewhat shy and would feel isolated in a larger project with separate apartments, and those who are especially sociable — two very different groups of people who may support each other well.

Conover believes the main advantage of the program is that the residents can remain independent while " . . . in a very sensitive way they can lean on the staff, on the programs, and on some other elderly people who would be just as happy to help out a neighbor."

Dr. Alexander Leaf of Harvard Medical School who recently spent two years visiting three regions in the world where there are especially high elderly populations, reports that in each case the elderly are held in high esteem, that even those over 100 perform essential duties and are used to prolonged labor. One of the more notable examples was a 130 year old woman, Mrs. Lasuria, who retired from her job as a tea picker just two years earlier. Perhaps more impressive are the examples of heart attacks going unnoticed by the victims living in these regions. The importance of physical activity — or lack of it — has been supported by a classic study of British postal workers. Those doing desk jobs suffered a higher incidence of and mortality from heart attacks than did their more active colleagues employed in mail delivery.

In contrast with the foregoing, consider the report by Beyer & Woods (1963) which shows that major portions of the American elderly population spend daily in excess of 30 percent of their leisure time watching T.V., and 20 percent of their leisure time napping and in idleness. Of a typical 15 hour waking day, wholly one-third (5 hours) is devoted by the elderly of this nation to grooming, cleaning, and meals (the remainder is considered leisure time.)

From An Elderly Center and Congregate Housing information pamphlet. □

Living In An Old House

Old houses aren't for everyone. Pipes may leak; space isn't laid out efficiently; wiring isn't adequate . . . the list of sensible reasons why one shouldn't buy an old house goes on and on. Yet there are growing numbers who would live nowhere else . . . .

Why do people endure the extra headaches of owning an old house — all for the privilege of living in a structure that sometimes behaves like a cantankerous spouse?

First, there is the romance of old houses. An old house is part of the collective memory of the human race . . . a living relic from the past. Long-forgotten joys and sadnesses linger in old rooms and on dark staircases. An old house continually reminds us that people have lived before. Through the house, we share an experience with those people from other times. Keeping up an old house is keeping faith with the past.

And an old house has character. In many ways it resembles a loving — but eccentric — grandparent. Having lived

Victorians

After the Civil War, America industrialized rapidly. New tools, new materials, new processes were suddenly available, and were quickly utilized by architects and builders. Power-driven mills turned out vast quantities of cheap lumber from America's virgin forests. Lathes and jigsaws turned wood into shapes never before possible on such a scale and the buildings today known as "Victorian" — often copies of Gothic, Renaissance, French or English styles — appeared in the cities and towns.

Today some of the finest buildings of the era remain; testimony to a time of expanding industrialism, buoyant hopes and inspired building. The drawings on these pages are from *Victorian Architecture — Two Pattern Books,* by A. J. Bicknell & William T. Comstock; *Woodward's Country Homes,* by George E. Woodward; and *Woodward's National Architect,* by George E. Woodward and Edward G. Thompson. See bibliography for details.

through many decades, an old house bears the imprint of all its previous occupants . . . acquiring a personality that is unique to that particular structure.

But despite imperfections, an old house usually exhibits an excellence of craftsmanship and detail that cannot be duplicated today. And while the creation of such extraordinary detail is beyond the ability of most contemporary workmen, it is within the ability of most homeowners to restore and preserve the work of the long-ago craftsmen.

Partly out of necessity, and partly out of a desire to develop their own manual skills, growing numbers of people are taking the do-it-yourself approach to restoring old houses. Taking care to keep the old details intact, they make the minor modifications required to keep an old structure suitable for modern living. And in the process, they discover the joy of living in a home finely made by the human hand. . . .

From *The 1977 Old-House Journal.* Reprinted by permission.

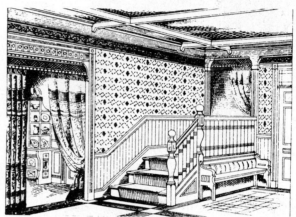

Historic Preservation

The National Trust for Historic Preservation is a private, non-profit organization chartered to encourage historic preservation in the U.S. Regional offices answer requests for assistance and information on preservation, neighborhood rehabilitation and home renovation. Typical projects involve individual houses, churches, public buildings, entire neighborhoods and districts, old commercial buildings and factories.

In certain cases, modest grants are made; one program features grants for young preservationists, such as a grant to a group of students and five advisers to purchase and restore a 115-year-old brick steamboat warehouse for use as a museum. The Preservation Press publishes the monthly newspaper *Preservation News* and the quarterly magazine *Historic Preservation*. For information: National Trust for Historic Preservation, 740-748 Jackson Place, N.W., Washington, D. C. 20006.

Publications on Restoration and Maintenance of Old Buildings (*see bibliography for further details*):
The Old House Journal, a monthly publication on "renovation and maintenance ideas for the antique house."
The Old House Journal Catalog, a buyer's guide to sources for products, materials and services for restoration of old houses.
The Old House Catalogue, 2500 products, services and suppliers for decorating and furnishing old houses.
How to Rehabilitate Abandoned Buildings by Donald R. Brann.
Built To Last: A Handbook on Recycling Old Buildings, Gene Bunnell.

Books on Victorian Architecture and Country Homes (*see bibliography*):
Victorian Architecture: Two Pattern Books by A. J. Bicknell and W. T. Comstock.
Woodward's Country Homes by George E. Woodward.
Woodward's National Architect by George E. Woodward and Edward G. Thompson. □

Ohmega
Salvage

by Vito San Joaquin

Ohmega Salvage is a demolition and salvage group in Berkeley, California. They learned the tricks of the trade from George Taylor (see Shelter, pp. 80-81) and have prepared the following article from their experiences.

The single most important principle to keep in mind in this business is that modern buildings are the product of mass production and have been put together with mass produced materials. The second is that "dismantling and salvage" (the term "wrecking" gives the wrong connotation!) is construction in reverse.

The constructor starts with neat stacks of mass-produced materials and assembles a structure — the dismantler or recycler reduces the structure back to neat stacks of reuseable material. Therefore, a tightly organized approach is necessary to make building dismantling on any scale practical.

Before starting any building, avoid pitfalls by doing some homework:

1. In choosing a building, first decide if it has enough material in it to be worthwhile — do a physical inventory of the amount of sheathing and the number of studs, rafters, and joists you'll get out.

2. Crawl under the building, get into the attic, punch holes in the walls — many structures hold nasty surprises (sometimes pleasant ones) like old fires that are covered over, termites, rot, and heavy linoleum on wood floors with non-water-soluble glues. Watch out for all these.

3. Figure out how much trash will have to be moved, where the closest dump is located, and what that will cost you in time, gas, and fees. Be sure, unless you are fortunate enough to have heavy equipment of your own, to steer clear of deals that involve removing heavy concrete footings and piers —their removal is heavy work and expensive!

4. *Analyze* the sequence in which your building was put together by the various craftsmen and then do the reverse: generally, we follow something of this order in dismantling: (1) windows, mirrors, electrical and plumbing fixtures, exposed pipe; (2) roofing, roof structure; (3) partitions; (4) exterior walls; (5) floors and sub floors; (6) joists and girders. Where possible, do each operation completely before starting a new one — it saves days in materials handling time and confusion.

Tips on building with used wood:
- *always look for the shortest length piece; this is a discipline you follow throughout the building process. It's a good feeling to cut a 10¼" block from an 11" piece of scrap.*
- *look at both ends of the board to be used. If one end is chewed up, make that the waste end.*
- *bevelling edges with small block plane makes used wood look good.*
- *remove all nails when dismantling. It will save your saw blades later on.*
- *discard any wood that has powder post beetle holes or dry rot (or cut off the rotted portion).*

The three most useful tools to the small building dismantler are the claw hammer (we prefer Stanley or Sears metal shank hammers); the stripping bar (George Taylor's and Ohmega Salvage's bars were specially forged and tempered and have a very *long*, relatively *thin* toe that will get under things);

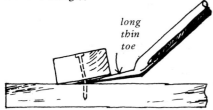

long
thin
toe

and the flat bed truck (preferably a 16-foot flat bed dump). This last (with plywood and stake sideboards) hauls and *dumps* your trash, saving countless hours of work. As a lumber hauler it can legally handle most of the longest material you will ever run into in the average building and you can dump your materials easily at your lot. For dumping lumber, you take a chain and chain binder around your load to keep it together.

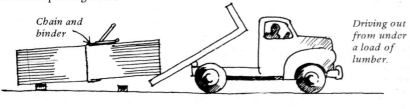

Chain and binder

Driving out from under a load of lumber.

Dump trash directly into truck below

Plywood chute to truck below

For safety's sake and to save time, remove trash regularly — but never let it get on the ground where you have to lift it back up onto your truck. For example, when taking off a roof, dump it from the eaves right into the truck below.

Trash within a building can be shoved out and down a plywood chute through a window opening that has been enlarged to floor level.

Finally, the basic task of the dismantler, apart from that of taking the structure apart, is the handling and moving of your material. Vast amounts of time and energy can be wasted in

repeatedly moving bulk items such as lumber — so have a well thought out plan to save effort. Try to establish a smooth flow of material from the point in the building where the lumber comes out, to the denailing horses, to the truck or position at the site where the material is to be stacked. Keep the area around the building clear, stack material carefully by size and type, and leave moving space between stacks. This latter applies both to the salvage site and to your storage lot.

For any lengthy storage of lumber, it is suggested that "stickers" or slats be placed in your lumber at regular intervals across the pile to keep everything from falling apart, to permit rained-on lumber to dry out, and to make easy calculations of how many pieces of a given kind of material you have. □

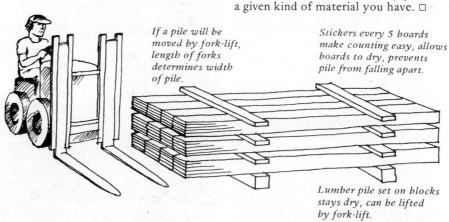

If a pile will be moved by fork-lift, length of forks determines width of pile.

Stickers every 5 boards make counting easy, allows boards to dry, prevents pile from falling apart.

Lumber pile set on blocks stays dry, can be lifted by fork-lift.

House Inspection

by George Hoffman

George Hoffman has inspected over 3,000 houses for prospective buyers in the San Francisco Bay Area, and has written an informative book on the subject, Don't Go Buy Appearances *(see bibliography). In the following article he summarizes some of the basic tenets of home inspection. Caveat Venditor.*

Inspecting a house properly takes two to three hours. If you can do it in less time, so much the better. But do it; it's essential. If you're buying, remember it's a used house. Check it as you would a used car. Don't take an agent's word for the condition of the house. Remember, agents are salesmen. If you own, it's important to give the house a yearly examination; like going to a doctor for a physical. Hopefully, you'll catch minor faults before they become serious in either case.

The tools I use are a steel ball ¾" in diameter, a long, thin screwdriver, a level, an electric tester, a flashlight and coveralls. Yes, you have to crawl beneath the house. In fact, that's the most important place to inspect. As for me, the house can appear first class above, but if it's rotting beneath, its fate is sealed.

When I approach a house I size it up. A glance tells me if it's been neglected. The condition of the paint doesn't bother me, although it does most people. I will admit that peeling paint doesn't make a house look good. But you don't need paint to preserve wood. Think of the 100 year old barns and fences still standing that never were painted. But if the paint is in poor condition, it has to do with house value, and that's what we're concerned with now. If it needs repainting, consider it in your offer.

What interests me when I see exterior neglect is dry rot — actually wood rot. Check any place where moisture can get into cracks: all around door and window trim, corners, and decks. Good caulking can prevent further rot. Window putty is important; if putty dries and shrinks, it no longer seals against water which runs down the glass, gets behind the putty and starts wood rot. Check window sash, especially where corners meet. That crack must be kept sealed on the exterior. Any window sash with metal braces indicates wood rotted joints. Test with a small screwdriver for soft, punky wood.

On the exterior I look at the roof gutters and for wood rot along the eaves. In snow areas especially, the moisture from melting snow can back up at the eave line and cause much wood damage. As I walk around the house I look for foundation cracks and termite tubes.

Roof. Try to learn the age of the roof. That will tell you its life expectancy. Identical materials in some areas of the U.S. will last longer than in other areas.

Durability of Roofing Materials:

Asphalt Shingles:	10 - 15 years
Tar and Gravel:	Average 15 years
Wood Shingles:	15 - 20 years
Wood Shakes:	18 - 25 years

Remember this: sunshine, not water, is the culprit that destroys roofs. A flat tar and gravel roof would last twice as long if you could keep water on it in the summer. Shade protects. Roofs on the north and east sides last longer than on the south and west sides. The sun dries the oil out of tar and gravel and asphalt shingle roofs, and they become blotters. If the asphalt shingles are brittle and bits of the colored mineral surface have come off, chances are the shingles are dry and lifeless. Tar and gravel roofs are the hardest to analyze: look for areas unprotected by gravel; these black areas should be smooth and flat. If sun-worn, the top course of tar paper shows small curled edges which are dry and soft, unlike tar paper which is jet black and firm. If you find many sun-worn areas, it's drying out beneath the gravel also.

Wood shingles, usually cedar, wear thin by sun and elements, crack, become brittle and blow off in a strong wind. Check the ridges: they often need replacing before the whole roof does.

There are many grades of shakes, but they all wear out. Shakes can be used on roofs with less slope than wood shingles; thus you'll often find wood rot on some shakes. Check carefully for that.

In this summary, I've mentioned only the most common roofs.

Now let's go inside.

Fireplace. Check if the damper operates freely and see if it closes tight to prevent hot air escaping when the fireplace is not in use. *Very important.* Jab at the fireback bricks to see if any are loose; if so, the grout should be replaced before bricks fall out. Test the grout; if it's crumbly, it's going to need re-grouting before long. Look at the lintel: is it strong or sagging, causing cracks in the grout? It won't get any better. If the fireplace has been used a great deal it is not a smoker. Who uses a smoky fireplace? If it smokes try adding three feet to the height of the stack.

Steel ball. This little tool is an extension of your eye. It will tell of sloping floors you can't see. Use it on floors near exterior walls to determine if the floor slopes from sinking foundation, termites or wood-rot-weakened lumber beneath. Near the interior wall the ball will tell you if there's insufficient support below. Make notes. When you crawl beneath the house, look for the causes of faults. If the floor sags in the middle of a room the joists span too much distance for their size, or posts under a beam have settled. You may have to stabilize a beam. The ball will tell you there's something wrong — it's up to you to learn what.

If there are wall to wall carpets, use the level on door and window frames, both horizontal and vertical, on walls, and if necessary on a long board placed on the carpet. Look for diagonal cracks. Check doors to see if they have been planed off to accommodate movement.

Don't be afraid of an unlevel house. If it has been stabilized but not leveled, you can live with it. I do. You can level your table legs or learn to serve plates with vegetables, up, gravy down.

Plumbing. Turn on all faucets, flush all toilets, check for leaks beneath counters and around toilets. Later you'll check for leaks beneath the house. Straddle the toilet and try to tip it from side to side. If it moves, there could be a broken toilet seal, letting water seep between the floor covering and the wood floor. Wood rot results.

As you turn on each faucet, note how much the volume of water decreases. If it gets down to a trickle, you no doubt have galvanized steel pipes with heavy mineral deposits on the inside: arteriosclerosis. It won't improve. No cure except re-plumbing. If fresh water pipes are copper, no deposits occur: quality.

Plastic drain pipes are approved in most areas. They're good, but should not be exposed to sunshine or will crack. Check the plastic drain pipe from the kitchen sink and the clothes washer to the main line. If it's a long run, it should be supported every six feet. Hot water softens plastic and it will sag without supports.

Look for clean-out plugs and caps. Wherever there's a right angle in drain pipes there should be a clean-out. With them maintenance is easy; without them you're short-changed.

If you have galvanized steel pipes for fresh water, check all the joints you can see. Threading pipe weakens it. If you find heavy deposits of rust at any joint, chances are the pipe is severely weakened. This could be a serious problem with all the fresh water plumbing.

Electric. Electricity is a complicated necessity. But you don't have to be an electrician to check for safe wiring and fusing. Briefly, here's what I check:

The average home should have a 100-amp service entrance; larger homes up to 200 amp. The main fuse or circuit breaker box should be labeled with the amperage capacity. Your utility company will tell you if the wires leading into the box are the correct size. If that's in order, turn off the power, remove the plate covering the fuses or circuit breakers. Get samples of number 14, 12, and 10 wire. Compare samples with each wire, red or black, that is connected to a terminal screw feeding the fuse or breaker (see Fig. 1). Number 14 wire takes a 15 amp fuse; number 12, 20 amp; number 10, 30 amp. You're checking for overfusing. Never mind the old saying, "Don't put in a larger fuse than you remove." It may be too large to start with. You're going to the source. Electricity is heat. Forcing too much current through small wire makes excessive heat which can start a fire. If you follow the fusing mentioned above you'll be safe.

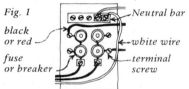

Fig. 1
black or red
fuse or breaker
Neutral bar
white wire
terminal screw

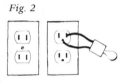

Fig. 2

Outlets in the kitchen and bath should be individually grounded. Figure 2 shows two types of receptacles. The one on the left is no longer installed, as it is not as safe as the other, but many are still in use. Many people are substituting the one on the right to update their kitchens, but they are not adding a third ground wire; without this it is no safer than the old outlet. To test if the modern one is properly grounded, insert one wire of an electric tester (Fig. 2) into the smaller of the two slots (hot side), the other wire into the ground (middle) hole. You should get as much current as when testing the two slots. If you don't get any current using the ground hole, try the longer slot and ground. If you get current you have cross polarity and even with the third wire your grounding is defeated. We can't go into any more detail in this writing.

Now turn off the electricity and remove the plate covering the receptacles in the kitchen. Check the wire size. Is it number 12? Good. If number 14 you are underwired. Kitchens demand much electricity, and require number 12.

The use of ordinary lamp cord in permanent installations is forbidden; never under carpets or hidden in walls. All splices in house wiring should be in junction boxes. Splices should be soldered and taped, or twisted together and covered with solderless connectors.

This is a very brief outline of electric inspection. But it's a start. Study a bit and learn more. Both knowledge of and the use of electricity will serve you well.

Tile. Ceramic tile is durable but the method of application is important. Tile glued to a wall, mastic method, is not as good as when applied by the mortar, or mud, method. With mortar they first put tarpaper on the wall, then wire, then a half inch of concrete. The tiles are pressed into the fresh concrete: not much chance of moisture penetrating to the wall. With mastic the tile is glued directly onto the wall. Hairline cracks in the grout can allow moisture to reach the wall and cause wood rot. In both instances, the wall should be made of waterproof material.

How can you tell which method of application has been used? Easy. Look at the edge.

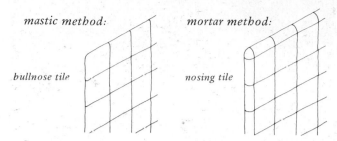

mastic method:

bullnose tile

mortar method:

nosing tile

Both have rounded edges, but if mastic, the tile itself has a round edge and is flat against the wall with only ¼" thickness. If mortar, the tile is ¾" from the wall, and has an extra round piece of tile covering the edge: this is best quality.

Heating. If the heating system is a forced air gas furnace, there are several important parts to check that are often ignored. Once a year the fan and motor need oiling, unless it's a sealed unit. The fan belt may have large cracks, in which case it will break on Christmas Eve. Replace it now. If the fan and motor are dirty it means the filter hasn't been changed often, perhaps never. Dust on the motor allows it to overheat. Dirty fan blades don't cut as much air; efficiency is lost. But most important is the heat exchanger, or firebox, which is hidden. It's the metal surrounding the flame. You can inspect the firebox by inserting a mirror into the hole leading to the burners. A crack, or pin holes in the firebox means carbon monoxide leaks into the plenum and is blown into the heating ducts. You have to have a new firebox. In most areas the utility company will check the heat exchanger at no charge.

Gravity furnaces also have heat exchangers. These should be checked, for they do burn out. Gravity furnaces have no motor, fan or filter.

With any gas furnace there should be oxygen venting: fresh air into the room. One square inch per 1,000 BTU is plenty. Vent pipes to carry away burned gas fumes must be checked for leaks. They rust out. Many floor furnaces have long runs of vent pipes beneath the floor. It may be a hard crawl, but those vents must be checked.

Drainage. Drainage is the most important part about a house. If too much moisture gets beneath a good foundation the house can settle. Beneath the house, look for indications of past moisture. Wet earth when dried leaves cracks like alligator hide. When standing water recedes, it leaves a mud line on concrete or wood posts: look carefully for that. In a full basement, look for water stains where floor meets wall. Stored goods on raised platforms indicate past trouble. Look at the platforms. Are the boards rotting? Boxes deteriorate with moisture. Smell for moldy conditions. Look sharply for manmade channels or ditches used to direct water out of the basement.

Wet earth in the crawl space beneath a house is an indication of trouble.

Admittedly, this has been a brief outline for house inspection. Mainly, though, I hope I've alerted you to the importance of inspections. Good luck. Keep a sharp eye. □

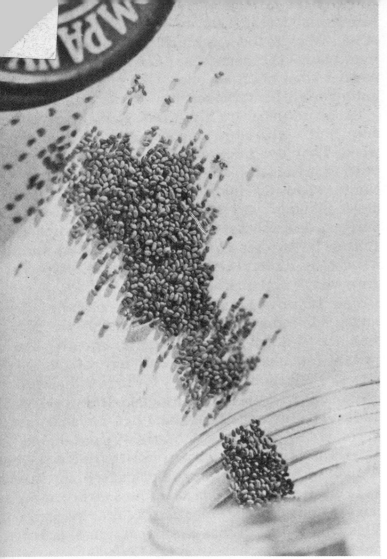

Drywood termite frass, tiny oval pellets pushed out of holes in the wood by the termites.

Termites

by George Hoffman

Damage to your house by termites and other wood destroying insects can be prevented by learning a few basic facts. Once you know their habits and how to detect them, you've got them under control.

Let's start with termites. There are three different kinds; each has different habits and conditions necessary for survival. *Subterranean termites* are the most common. They live in the earth, but the workers travel back and forth from the house to their nest carrying bits of studs and joists. Two important points: To survive, they must return to the earth frequently for moisture. And they do not expose themselves to air. To protect themselves from their enemies (ants mainly), termites build shelter tubes on the side of foundations or anything else handy, so they can get from the earth to the wood. These passageways, made of earth and body wastes, are about ¼" wide, brown in color. Their length depends upon the persistence of the termite and the scarcity of food. Ordinarily, termites won't build the tubes more than ten or twelve inches high, but I have seen them as high as six feet.

The most important lesson in termite control is: *no earth-wood contact*. If wood is touching the earth, no tubes are necessary. The termites could be hollowing out the wood and you wouldn't know it until the structure weakened.

Keeping in mind the points you have just learned, let's examine your house for termites. First, walk around the perimeter of the structure, exposing the foundation and searching for tubes. Look carefully behind bushes. If earth is touching wood it's suspect. Remove the earth and look carefully for the insects. They are grayish-white, about an eighth of an inch long, with a light brown head. When disturbed, they move slowly as they try to find their bearings.

Don't panic if you find tubes or termites. After you clear away the earth or the tubes, the termites in the wood will die because you have denied their access to moisture. To prevent reinfestation, you have a choice. Poison the earth around the foundation with an approved chemical, such as pentachloriphenol or chlordane, or make certain there is no earth-wood contact and inspect your house once a year for tubes. No tubes, no termites. I don't use the poison but I make certain I have no earth-wood contact and I check periodically for tubes. If only two inches of concrete separates earth from wood, you can examine the exterior quickly.

Now you've only inspected half your house; the worst is yet to come. Conditions are better for termites in the crawl space beneath your floor — dark, damp, unmolested. Use a strong flashlight and don old clothes to crawl beneath the floor, searching for tubes. Sometimes the tubes will be sticking straight up from the earth like an asparagus patch as they reach for wood. Look carefully around furnaces because the insects like warmth. Any loose piece of wood lying on the earth is an invitation: food to keep a colony on your property. *Do not store wood on earth beneath your house.* If your floor joists are too close to earth for you to crawl under, you can cut tunnels to get to those inaccessible areas for a better look. Or consider having the area chemically treated for prevention. Any tubes that are found should be cleared away and the earth treated.

We mentioned earlier that subterranean termites do not expose themselves. The one exception is the reproductive form. They do swarm, coming out of the earth on warm, humid spring days in their nuptial flight. They do not destroy wood. The reproductives have yellow-brown to black bodies with two pairs of equal-sized wings. Their bodies are chemise shaped — less elegant than the hourglass-waisted flying ants.

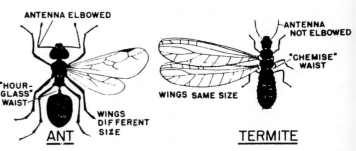

Note different wing size and bent antennae on ant.

The two are frequently mistaken for each other. If you discover flying termites or discarded wings on your window ledge, don't worry. Should they be fortunate enough to mate, they still must make earth contact to survive and you can prevent that. However, the presence of wings indicates there is a colony nearby, perhaps on your property.

Drywood termites have different habits from their subterranean counterparts. They live in the wood, but they don't do as much damage. Their geographical area of operation is in the warm to hot climates of the U.S. To detect infestation, look for tiny, football-shaped pellets on window ledges, interior and exterior, baseboards, shelving or any flat surface. These pellets, called *frass*, are pushed out of small, usually oval-shaped holes. Attics should be carefully checked. Damage by these insects can be recognized by galleries cut across the grain of the wood; any wood. These termites aren't fussy; they like oak, redwood, pine and fir. Control of drywood termites is often accomplished by injecting an insecticide (such as Paris green) into the galleries. If there is a general infestation the house may have to be fumigated: not a do-it-yourself job. Prevention is difficult. Fitting tight screens on attic ventilators, treating wood, sealing cracks and end grains is about the best you can do. Periodic inspection and treatment maintains control.

Dampwood termites are larger than the two already mentioned. They live primarily in decaying wood which must have moisture to make it decay. So keep moisture away from wood. A leaky toilet seal or other plumbing is a frequent cause of dampwood termite infestation. Inspect closely beneath bathrooms at least once a year. And check the eaves of a house. Wet, rotted wood is an excellent condition to attract dampwood termites.

We are constantly warned to be on the watch for termites or they will terminate our houses, but not so much is said about *powder post beetles*. These beetles do as much damage as termites. Their prevention is difficult but their detection

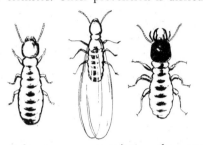

Subterranean termites: the mature worker, the winged reproductive, the mature soldier; their predator, the ant.

is easy. Beetles make tiny holes in wood. They look like someone has been playing a dart game on your floor joists. The substructure of your house is the most common area of infestation. When you check for termites, look for beetle holes. If you find holes it's evidence that the beetle larvae have been in the wood for a year or more, eating and damaging the wood. Not until the larvae reach the adult stage do they make those emergence holes to crawl out and start another colony. But worse luck, some larvae stay in the wood. Where you see an infestation, strike the wood and watch for frass to fall out in the form of fine, white powder. Test the wood for damage by jabbing at it with a screwdriver. If you can't detect frass, chances are the beetle is not active. To make sure, I apply creosote to the surface. This turns the wood dark. In six months I look again. Any activity will deposit white powder on the black surface. No powder, no beetles. If there is powder I brush with Penta preservative, then check again in a year. A word of warning: should you find that most of the substructure lumber is infested, fumigation is the best answer because the gas gets into cracks and crevices you can't reach. The gas kills but supposedly leaves no residual protection.

The beetles love old furniture. Hardwood flooring and redwood is just as inviting as oak and fir. Storing fireplace wood in the house often results in beetle infestation. How many of you have picked up a piece of firewood and seen tiny holes in it with fine powder falling out? Beware of beetles; burn the wood.

I left the worst for last. *Dry rot*, which is wood rot, has nothing dry about it. It is a fungus which can only grow if it has moisture and one other condition: no air circulation. Leave a piece of wood on a concrete sidewalk and the fungus will start growing only where the wood touches the concrete. Eventually the fungus will penetrate the entire piece of wood, but it will take years. Remember: moisture and no air. Look at a deck. The moisture gets between the decking

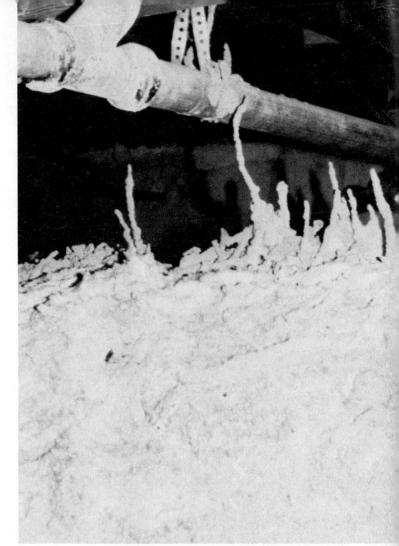

Termite tubes connect earth to wood

and the joists. That's where wood rot grows. A leaky toilet spreads moisture under the floor covering and wood rot results. Spilled water around bath tubs and showers, or beneath sinks is a common cause of wood rot. Beneath a house, poor ventilation and a severe moisture condition promote fungus which destroys wood. A leaky roof can rot the sheathing beneath tar paper. But you're really asking for trouble by allowing wood contact with soil. Damp earth, no air = wood rot. Fence posts; if you must have wood in the earth, treating it with a good wood preservative beforehand will help. Redwood is more resistant to rot than fir or pine, but it does rot.

There is a good product on the market for repairing minor wood rot. Boat owners have been using it for years, and they have had more experience with rot than most people. Calignum, Git-rot and Cure-rot are three such names. Ship chandlers carry the product as well as some building supply houses. It's a plastic product that is inserted into rotted areas. Used properly it penetrates the infected wood where it hardens and becomes stronger than wood. It is especially good on window and door frames, deck railings and deck joists. So don't replace any rotted wood until you have investigated this material.

Keeping in mind the information you have just learned won't make you a pest control operator, but it can give you peace of mind if you apply the knowledge to your house. Let the insects have the fallen wood in the forests, not the joists in your house.

For further information there are two good pamphlets on wood destroying insects. U.S. Dept. of Agriculture, Forest Service Publication 1284; U.S.D.A. Home and Garden Bulletin 64. □

Industrialized Housing

Domes

Part of any experimental process is observation of results and evaluation of achievements. Shelter Publications began almost ten years ago as an experimental housing group, dedicated to exploration of new building techniques and publication of the best available information on experimental building. Over a two year period, 1970-71, 17 domes were built at Pacific High School, near Saratoga, California. In all, 10 different *type* domes were built and observed. The work at Pacific (much of it admittedly unprofessional and crude) and input from a growing network of contributors produced first *Domebook One* in 1970, then *Domebook 2* in 1971.

Domebook 2 was in print for about four years, during which time literally hundreds of dome builders wrote us of their experiences and observations. As time went on, as we correlated other domebuilders' experiences with our own, we came to believe that domes had unique and specific drawbacks that made them — in our opinion — less efficient and practical than conventional buildings, and we decided to cease publication of *Domebook 2*.

Panaceas have always had great attraction in America. Domes fit neatly in this category. They appear exciting and revolutionary, they promise untold advantages, the simple geometrical aspects have great appeal, and moreover, they photograph well.

On the following four pages are some aspects of domes and dome building generally overlooked in books and articles on the subject. Here is the other side of the picture (we admit to being biased), points to consider along with the positive claims being made in current literature and by dome enthusiasts such as R. Buckminster Fuller.

If a person wants to live in a dome it can be done. But a dome is certainly not a panacea, and a balanced analysis of the drawbacks as well as the merits would seem advisable to anyone contemplating a dome home.

'. . . a truly unprecedented and advanced work is not that which uses superficial brilliance to make a temporary and sensational impact, or that which seeks to take one by surprise by means of ostentatious, acrobatic contortions, based on momentary 'finds,' but only that which is justified by a continuing, living tradition, that which endures because it is put to the test again and again, within each new context . . .'

Elements for Self-knowledge
Aris Konstantinidis

Domes and Conventional Construction: A Cost Comparison

The primary advantage of domes is generally said to be lower building costs. Yet even though a dome may use less materials to cover a given amount of space, these materials must be of higher grade, and are more expensive. To make a comparison, Wayne Cartwright has calculated the materials and costs for the most widely-sold dome in the U.S. today with a stud-frame house of the same floor area:

The dome is a 40 ft. diameter hemisphere, built in the "Pease Dome" or "Cathedralite" style. The triangles are approximately 9 ft. by 8 ft., framed with kiln-dried Douglas Fir 2 x 4's, sheathed with ½" c.c. No. 1 ext. plywood. Insulation is R-19 Technifoam. ½" sheetrock is on the inside; the dome exterior is completely covered with asphalt tab shingles. Floor area is 1,256 sq. ft.

The conventional structure has the same floor space and same volume as the 40' dia. dome, and more surface area. There is space for a full height second

Domes/Rectangles

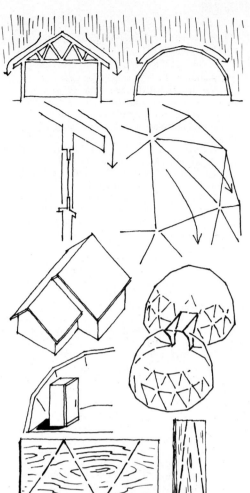

1. A rectangular structure is built of walls, which are vertical, and a roof. The roof acts as an umbrella, keeping most (not wind-blown) moisture off the walls, windows and doors. A dome is all roof; water, including mists or fog, pours over the entire surface. Unless the dome is covered with shingles, the slightest pinhole causes leakage.

2. Rectangular buildings are shaped by available materials — wood, stone, adobe, etc. — and the laws of gravity. Domes are shaped by polyhedral geometry and materials must be forced to carry out the abstract concept.

3. The dome framework, due to its tightness, is continually under stress. As temperatures change it expands and contracts. It is always working, always straining at the seams.

4. Domes must be built of higher grade materials. The kiln-dried lumber required for framing is over twice as expensive as construction grade lumber used in stud framing. See cost comparison below.

5. Almost all building materials come in rectangular shapes. They must be altered to fit polyhedral shapes, either with resultant waste or more complicated cuts.

Also, once materials are cut for dome assembly they are difficult to recycle in another building.

6. A far greater variety of materials can be used in conventional construction: rock, adobe, used wood, doors and windows, construction grade lumber, etc.

7. An important feature of an owner-built home is the possibility for later expansion. With a perpendicular wall, you merely add on more roof and walls, all at 90º. With a dome, however, you weaken the structure by cutting into it and must cut compound angles and tie into multiple facets when adding on.

8. Similarly, constructing interior partitions in a dome is far more time-consuming, due to the compound angles.

9. The dome's well-publicized "more space for less materials" actually means more cubic area (overhead) that is hard to utilize and must be heated.

10. *We* are vertical to the earth. So are refrigerators, beds, bureaus, tables, kitchen counters, etc. These things fit best in a rectangular space, less efficiently in circular space.

11. Each triangular facet of a dome faces center; this magnifies noise. Also, smells circulate throughout the entire dome.

---

story in the higher portion of the house. Walls are framed with construction grade Douglas Fir 2 x 6's on 2' centers, sheathed with V-rustic clear redwood 1 x 6, braced with 1 x 6 Doug. Fir, and blocked horizontally on 4' centers. Insulation is R-19 fiberglass, and inside is sheathed with ½" sheetrock. The roof is framed with double fink trusses:

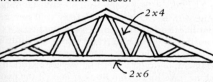

These allow the interior partitions to be located randomly as in the dome. The ceiling is insulated with R-19 fiberglass and panelled with ½" sheetrock. The roof is decked with 1 x 6 Douglas Fir and covered with asphalt tab shingles.

Materials required for each structure, not including floors, doors or windows and average prices (in San Francisco Bay Area) are as follows:

Pease dome

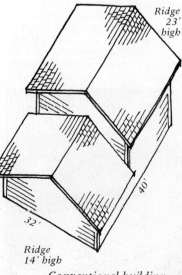

Ridge 23' high

Ridge 14' high

Conventional building

Comparison Chart
40' dome
1256 sq. ft. floor area; 2512 sq. ft. surface area.

Frame: 2x4 kiln-dried D.F.	3,558 lin. ft.	$2,740
½" c.c. ext. grade plywood	78.5 sheets	1,030
R-19 Technifoam insulation	78.5 sheets	1,275
asphalt tab shingles	2,512 sq. ft.	730
½" sheetrock	78.5 sheets	313
	TOTAL	$6,088

Rectangular building
1256 sq. ft. floor area; 2316 sq. ft. wall area; 1548 sq. ft. roof area.

2x6 D.F. const.	2,704 lin. ft.	$ 960
1x6 " " bracing	132 lin. ft.	17
2x4 " "	1,100 lin. ft.	209
1x6 " " roof sheathing	3,715 lin. ft.	483
1x6 V-rustic clear redwood	5,558 lin. ft.	2,112
R-19 fiberglass	3,572 sq. ft.	893
1x6 clear redwood trim (facias, soffits, & rakes)	618 lin. ft.	420
asphalt tab shingles	1,548 sq. ft.	450
½" sheetrock	111 sheets	445
	TOTAL	$5,989

Notes:
1. R-19 fiberglass and R-19 Technifoam have equal insulating properties.
2. Clear redwood siding on conventional house will last over 100 years whereas shingles on both conventional house roof and entire dome surface must be replaced at least every 10 years.
3. If cedar instead of asphalt shingles were used on both structures, the dome would cost $572 more.
4. Bill Woods' Dyna Domes use considerably less lumber in framing than Cathedralite domes.

Dome Letters

Friends!

Regarding your disillusion with domes, unfortunately I find myself in sympathy — unfortunate in that I'm already locked ($) into construction of one. I expect we'll be living in the thing within a year, which will be but a beginning. Your publication *Shelter* says a lot that I have been feeling but hadn't verbalized. Thanks! Anyway, here's a description of what I've got going — Plans are for a cluster of 1 big, 2 or 3 smaller domes — time approximately 5 years — $ and time permitting. Got the floor, frame, plywood, and building paper for the main dome up last fall and left it to itself for the winter. A few leaks but held the snow load fine. Am in the process of roofing with asphalt shingles. Dome is 35' dia., 11/20 sphere, 5 freq., icosa., class 1, method 1 (Domebook II). When first started expectations (based on 12' model of frame) of spaciousness were eagerly held — and the regularity was/is something to behold. But appreciating in your head and functioning with the structure are different processes! The space is there alright, but we're not a family that wants one big room for a dwelling. The dome doesn't lend itself to internal subdivision very readily without loss of characteristics which make domes so (romantically) appealing. What I've got to do is impose my own internal/ external irregularities upon the purity of the structure so as to allow the finished dwelling to be both mechanically sound and aesthetically happy. No small task and I pray for the patience to do it right! . . . Domes are not "where it's at" but I'm gonna have one.

Peace
D. Mitchell Freeman
Chewelah, Washington

To Shelter:

I've just finished reading the 32 page section on domes in your book. It was a little disheartening for me to read because next week my husband and I are moving into a dome.

A little background — Like many other people we moved out of the city and bought a small piece of land. Our goal was to live a different kind of life and let it happen more or less naturally. I found that it's difficult to let go of "old ways" and I'm still working on making those changes in myself. In terms of shelter, the "old ways" meant electricity, hot running water in the house and a gas range for cooking. My husband's dream was to build a dome and he had drawn elaborate plans for everything. Much of the excitement and information coming from Domebooks. But we found that we couldn't build such a big and expensive structure right away so I went to work in town and Neil built a small temporary house.

That small temporary house turned out to be just the kind of building that you are now advocating or looking at. The house is the shape of the level land it sits on. Most materials are old wood scrounged, or the lowest grade of what someone had to sell. There is a huge hand built stone (from down the road) fire place in the center. The southern face of the house is almost total glass brought from Neil's grandfather's house in L.A. In short the house is a beautiful, livable little structure which I've grown to love.

Since the little house was finished — it's not finished but we have lived here for 1½ years — Neil has gone on to build the dome. Since we now have a 6 week old daughter and no income, we are moving to the dome and are renting our little house. I have such sad feelings to leave here but it must be done.

After reading your article I can imagine the effect that it had on Neil and feel a real need somehow to let you know that a dome can be a beautiful space and *work*. The dome is wood construction and shingled with wood. It has very few leakage problems and has survived a rainy Oregon winter quite well. Neil has not confined himself to a one room dome but has built on a big 32' dome (2 story) and a small 24 ft. dome connected with a square 3 story tower. Attached to the large dome is a two story addition that is polygonal. It's truly an amazing structure. It looks like a Hobbit or fairy castle from a distance. Inside is totally unfinished so I can't comment but I think it will be lovely.

I think the big difference between our dome and many other people's that aren't successful is that my husband is a terrifically talented person. He is sort of a jack-of-all trades with this creative imagination thrown in. Most people that build domes perhaps let themselves be bound by the mathematics and the confines of the dome where Neil just added on and improvised anyway.

I think most of what was said about the dome's drawbacks are very true. At the same time the problems can be worked out somehow. . . .

I think it would be excellent to have a successful improvised dome in your next book.

Sylvia Stewart
Applegate, Oregon

People:

Hi! I'm Jack, and after reading your articles in *Shelter*, I feel I owe you an apology. From the tone of your comments re. *Dome Book II*, apparently the only people who bothered writing to you were the ones who tried to build a dome and screwed it up and were pissed off.

Maybe I'm too late to stop a bunch of bad vibes, but I read *Dome Book II* and read it and read it and read it, and finally screwed up the courage to try to build my own home. When I started, I knew which end of a hammer you hit the nail with, but that was about it . . . .

Sure, I probably could have built a conventional frame house cheaper. We were up to about $1200 by the time we had us inside and the rain outside, but I wouldn't have.

Anyway, about 18 months ago we put up a 24', 5/8 - 3 freq. dome and it's still standing. Page 138 & 139 of *Shelter* pretty well sum up our likes and dislikes concerning the house except for one or two critical likes. We built it ourselves in the face of an organized opposition but we didn't feel like we were working alone. We had *Dome Book II* and all the people who contributed to it to lean on. Thanks to your help and guidance, it's up, it's snug and it's home.

. . . Let me make one more comment. Don't feel bad about really getting into domes and trying to push the idea, you got at least one family out of the rent racket without leading them into the mortgage hassle. You will always have our love and support.

Jack Silvia & Family
Little Compton, R. I.

Dear Sir:

I recently ran across a copy of your *Dome Book II* and was fascinated by the contents.

About 5 years ago I got a hold of an old Pease (22' x 26') dome which had been used as a drivcup hamburg joint. My family pitched in and we took it apart — moved it 2 miles up a small lake by boat — and put it back together; after clearing out trees and building a platform.

This building hasn't allowed nearly as much creativity, uniqueness, and individualism as those shelters you have written about. We all thoroughly enjoyed the project though. It has made me want to go on and do more building . . . .

Sincerely,
Stephen W. House
Brookline, Mass.

Hi,

For the past three years we've been living in our dome-home in Killaloe, Ontario (28' hemispheric, 4-phase alternate). I've just read the "smart but not wise" comments in the second edition of Domebook II and the thoughts and experiences are an *identical* reflection of my own.

Our dome was built using scrap wood and plastic. The plastic hubs have broken, the vinyl windows have all but disintegrated, the plastic sealing cement (on canvas for the covering) has broken down and no longer holds out the weather, and the styrofoam insulation is still in tact but the mice are beginning to slowly eat it away. At this point the moisture inside is slowly destroying our possessions. I can't say I was completely disillusioned by the breakdown of the plastics but they sure were of a lot shorter life than I expected. To make a long story short, I must rebuild, and I have been looking forward to your promised edition of "Shelter" for inspirational information.

On our farm we have an abundant supply of rock, dirt, and wood and I'm hoping to combine these natural materials to construct a simplistic but effective shelter attached to the dome if possible, as well as repairing the dome itself . . . .

Regards,
Brian Barstead
Lynwood, Washington

Dear Sir,

I am a journeyman carpenter, came to Hawaii to build a dome by the ocean. I mean the (proper) way — plans, codes, permits, the whole trip. What a laugh. I'm hung up in a world of so called freaks. Long hair & dope is the extent of their freak out. I'm originally from Detroit where I was a machine designer. I just can't go back to that life. I feel as though a dome is the only thing I must do right now. People living in self sustained dwellings is the only way we will ever regain our freedom. Even the word freedom says it (FREE DOME) — well, add an *e* and it does. This thinking has cost me a wife and many friends but as long as I see (Whole Earth) (Domebook I & II) I don't feel so alone . . . .

Sincerely,
Larry Fatalski
Honolulu, Hawaii

Domes on Venus

Have *Domebooks No. 1 & 2* and *Shelter,* totally invaluable. Proper domes will survive the change of the earth (Earth) from the third to the fourth dimension when the dense & 90 degree structures will fail.

The (most) buildings on one dimension of Venus are low density hemispheres of a cellular material much stronger than our plastics — they are semi-translucent. It is into this vibration an event in the making will project the Earth. In our time . . . .

John Eversole
Santa Barbara, California

Do It Again? Never!

. . . After four years, the cat, dog, five children and Chris and I are happily living in the completed dome. Inevitably, guests walk in, look all around — up at the ceiling, the huge windows — and start asking questions. "What are some of the problems that you have encountered with the dome?"

The acoustics have negative as well as positive aspects. Vivaldi is glorious, but for the first two years our bedroom was in the loft, and a ten-year-old child eating Rice Krispies at 7 a.m. could bounce us out of bed. If a whisper could be heard in the kitchen, it was a shout in the loft. The only solution was to give the loft to our ten-year-old daughter, who sleeps like a log, and build an attached master bedroom and bath. Ah, peace and quiet.

"How much did the dome cost?" About as much per square foot as any custom-built home.

"Does your dome leak?" No. We used the most conservative roofing techniques possible, and the structure is not only leakproof, but it is very efficient to heat. Every exterior joint was flashed with galvanized metal, and cedar shakes cover the dome with a beauty that blends with the surrounding terrain.

"Would you build a dome now if you had to start all over again?" No. The initial shell was completed in one day, but the finish work was incredibly difficult and time-consuming. I would recommend domes to skilled carpenters or cabinet makers; but for ten-thumbed amateurs or people contracting to have the job done, it is simply a very difficult or very expensive project. There were literally weeks before a right angle could be found. Sharp little acute angles joined to form pentagons or hexagons.

We finished the interior with rough-sawn redwood and it is beautiful, but each triangle had to be measured, re-measured, beveled, glued and, finally, precisely nailed with rust-proof finishing nails. Our cedar shake roof had to be cut and fit to match the constantly changing angles of roof-line. Each angle had to be flashed with galvanized metal to guarantee a leak-free roof. Each window was custom-made to accentuate the integration of the dome with the surrounding trees and shrubs; and they are works of art, but expensive. It is a beautiful creation, and we dote on it like misguided grandparents; but do it again? Never!

Robert Evans
California Living Magazine
San Francisco Sunday
Examiner & Chronicle

Dear Shelter,

As you probably know, the Biosphere (U.S. pavilion at Expo '67) has burned up. This fire, even if it made a hole in our hearts, tends to prove what you were saying in previous publications. Plastics tend to be short lived and very dangerous with fire. In this case, all acrylic panels melted in less than 15 minutes. The fire was caused by a welder working on the dome. This was the third fire caused by welders there (the two others were resolved without troubles).

The most surprising declaration was from the architect who built the dome in collaboration with Fuller: he said they had taken special measures against fire in the planning. At that time they feared terrorist acts (U.S. is often synonym of imperialism here). Now we can ask ourselves, are those people really credible? What would have happened if the fire had occurred in a dome covering a town as Fuller suggests?

This fire made many Montrealers sad because the dome was a kind of structural symbol of "man and his world." Since the dome frame is still intact the city is planning to transform it into a kind of outdoor gathering place where they can present various spectacles.

Salut,
Guy Huot
Student in Ecole d'Architecture,
Université Laval, Montreal

Foam Domes

Adapted from an Article by Suha Ozkan

When a disastrous earthquake struck Gediz, Turkey in 1970, 405 polyurethane foam plastic domes were contributed to homeless residents by a German chemical firm. The crew, chemicals and machinery were all flown in from Germany. The domes were being produced *in situ* within two days of the disaster. Two inflatable positive molds were used; after spraying, the plastic was allowed to harden for 30 minutes, then removed. Doors and windows were then cut in, and the domes were carried to their sites by 10-15 men.

Rapid erection of the foam domes allowed the government time enough so that permanent housing did not have to be rushed in Gediz, as it was in other areas hit by the disaster.

With permanent homes now completed, the domes (which are disintegrating) are still used for living by a few people, or as extra room for animals and storage.

Families added porch-like additions to their domes, as kitchen or storage space and for windbreaks. Sometimes three or four domes were connected together if owners were of the same family.

Interviews with the residents of Gediz have produced the following conclusions on the performance aspects of the plastic foam domes:

Positive aspects:

1. Insulation was excellent; fuel was reduced to one quarter of regular consumption.

2. Construction was rapid; this gave the government agencies and inhabitants more time to design and erect permanent housing.

3. The domes were portable and easily moved.

Negative aspects:

1. The hemispherical shape created difficulties in furnishing, did not allow cellular growth, and discouraged additions.

2. The plastic foam was almost impossible to repair within the resources of the community. Asphalt, gypsum, cement, even sewing with wire were tried, with little success.

3. Ventilation is poor. In summer the domes are insufferably hot. People reported that the plastic made them dizzy; that it exuded a bad smell in summer; that restless sleeping was common due to gradual suffocation.

4. The domes break easily when transported. Once broken, the pieces are not reuseable other than as filling material.

5. There is insufficient light inside.

6. Once the protective outer layer was washed off, a previously unknown black vermin came into existence. This vermin formed colonies within the granular structure of the polyurethane. □

Polyurethane Foam

In the 1960's, polyurethane foam was regarded as a significant breakthrough in the building materials industry. It was a better insulating material than fiberglass, it could be shot from guns and would adhere to almost any surface (entire experimental buildings with curved surfaces were being constructed of the material), and it was advertised as "self-extinguishing" by plastics manufacturers.

From *Fire*, ABC news special, 1973:
Wastebasket with empty milk containers lit in corner of small room lined with polyurethane foam wall boards.

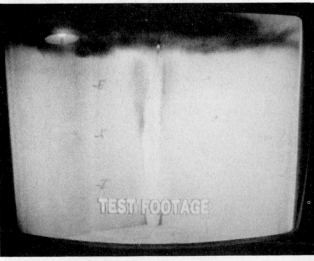

The foam — advertised as "fire-retardant, non-burning and self-extinguishing" by the plastics industry — is now on fire and flames have reached the ceiling.

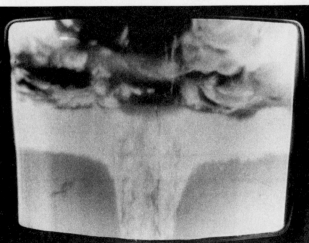

One minute, 40 seconds after the fire is lit, the room is engulfed in flames and poisonous gases.

In 1967 two children died when unprotected interior foam insulation caught fire in a garage where they were playing. "There was just no chance of getting to them in the garage at all," said their father, Jerry Childers. "When the insulation was burning, the flames were so red hot you could hear the fire roar. It sounded like a train going by."

Even after this tragedy and several disastrous large-scale commercial fires occurred, the plastics industry continued to advertise and market foams as self-extinguishing.

In May 1973, the Federal Trade Commission accused 26 plastics companies of falsely and deceptively advertising polyurethane and polystyrene foams as being non-burning or self extinguishing. In September of that year, a billion dollar class action suit was filed against five plastics companies, charging them with misrepresentation in selling highly flammable foam for uncoated interior uses that could result in low order explosions. In November 1973, an ABC-TV news special, *Fire*, covered the fire hazards of polurethane foams.

Foam had been tested in the Underwriters Laboratory tunnel tests, where a draft is blown across the ignited material being tested. Whereas a wood fire will increase as you blow on it, plastic material will often go out when a draft strikes it. In a more realistic test of what could happen in a home, ABC showed a small room built with polyurethane wall boards with a 25 flame spread rating (considered safe). A wastebasket with 12 empty milk containers was ignited in the corner of the room. In less than two minutes, the room had exploded in flame (see at left).

ABC news revealed that as early as 1969 the plastics companies knew of the hazard; they had been warned by Underwriters' Laboratories that ". . . the fire behavior of polyurethane foam with adequate oxygen supply is one of a very high flame spread, high early heat output rate, and production of large quantities of dense black smoke. . . ."

In 1974 a fire in a particle accelerator facility spread 430 feet in eight minutes after a workman ignited the foam with an oxyacetylene torch.

Ironically, flame inhibitors added to polyurethane foam usually produce toxic gases, or greatly increase the amount of smoke. Foam is made of oil and will burn like napalm once it gets started. If foam is used, fire officials recommend it be covered with a thermal barrier, such as 5/8" sheetrock or equivalent thickness of plaster.□

Sunbeams
From
Cucumbers

Gulliver's Travels *was written by Jonathan Swift in 1726. In his journeys to different lands, Gulliver visits the grand academy of Lagado, where:*

. . . the professors contrive new rules and methods of agriculture and building, and new instruments and tools for all trades and manufactures; whereby, as they undertake, one man shall do the work of ten: a palace may be built in a week, of materials so durable, as to last for ever without repairing. All the fruits of the earth shall come to maturity, at whatever season we think fit to chuse, and encrease an hundred fold more than they do at present; with innumerable other happy proposals. The only inconvenience is, that none of these projects are yet brought to perfection; and, in the mean time, the whole country lies miserably waste; the houses in ruins . . . .

. . . The first man I saw, was of a meagre aspect, with sooty hands and face; his hair and beard long, ragged and singed in several places. His cloaths, shirt, and skin, were all of the same colour: he had been eight years upon a project for extracting sun-beams out of cucumbers: which were to be put into vials, hermetically sealed, and let out to warm the air, in raw inclement summers. He told me he did not doubt, in eight years more that he should be able to supply the governor's gardens with sun-shine at a reasonable rate; but, he complained that his stock was low, and intreated me to give him something as an encouragement to ingenuity, especially since this had been a very dear season for cucumbers: I made him a small present, for my lord had furnished me with money on purpose; because he knew their practice of begging from all who go to see them . . . .

continued

1726 Gulliver's Travels: . . . *I turned back and perceived a vast opake body between me and the sun . . .*

The flying, or floating island, is exactly circular; its diameter seven thousand eight hundred and thirty seven yards, or about four miles and an half . . . it seemed to be about two miles high . . .

three hundred yards thick. The bottom, or under surface, which appears to those who view it from below, is one even regular plate of adamant, shooting up to the height of about two hundred yards. Above it lye the several minerals in their usual order; and over all is a coat of rich mould ten or twelve foot deep. The declivity of the upper surface, from the circumference to the center, is the natural cause why all the dews and rains which fall upon the island, are conveyed in small rivulets towards the middle, where they are emptied into four large basons, each of about half a mile in circuit, and two hundred yards distant from the center. From these basons the water is continually exhaled by the sun in the day-time, which effectually prevents their overflowing. Besides, as it is in the power of the monarch to raise the island above the region of clouds and vapours, he can prevent the falling of dews and rains whenever he pleaseth. For the highest clouds cannot rise above two miles, as naturalists agree, at least they were never known to do so in that country.

At the center of the island there is a chasm about fifty yards in diameter, from whence the astronomers descend into a large dome, which is therefore called 'Flandona Gagnole,' or the 'Astronomers Cave'; situated at the depth of an hundred yards beneath the upper surface of the adamant. In this cave are twenty lamps continually burning, which from the reflection of the adamant cast a strong light into every part. The place is stored with great variety of sextants, quadrants, telescopes, astrolabes, and other astronomical instruments . . . .

Jonathan Swift:
Gulliver's Travels, 1726

. . . Yet, I found my self so listless and desponding, that I had not the heart to rise; and before I could get spirits enough to creep out of my cave, the day was far advanced. I walked a while among the rocks; the sky was perfectly clear, and the sun so hot, that I was forced to turn my face from it; when all on a sudden it became obscured, as I thought, in a manner very different from what happens by the interposition of a cloud. I turned back, and perceived a vast opake body between me and the sun, moving forwards towards the island: it seemed to be about two miles high, and hid the sun six or seven minutes, but I did not observe the air to be much colder, or the sky more darkned, than if I had stood under the shade of a mountain. As it approached nearer over the place where I was, it appeared to be a firm substance, the bottom flat, smooth, and shining very bright from the reflexion of the sea below. I stood upon a height about two hundred yards from the shoar, and saw this vast body descending almost to a parallel with me, at less than an English mile distance. I took out my pocket-perspective, and could plainly discover numbers of people moving up and down the sides of it, which appeared to be sloping, but what those people were doing, I was not able to distinguish . . . .

The flying, or floating island, is exactly circular; its diameter seven thousand eight hundred and thirty seven yards, or about four miles and an half, and consequently contains ten thousand acres. It is

Space Colonies

... The human race stands now on the threshold of a new frontier, whose richness surpasses a thousand fold that of the new western world of five hundred years ago.

That frontier can be exploited for all of humanity, and its ultimate extent is a land area many thousands of times that of the entire Earth. As little as ten years ago we lacked the technical capability to exploit that frontier. Now we have that capability, and if we have the willpower to use it we can not only benefit all humankind, but also spare our threatened planet and permit its recovery from the ravages of the industrial revolution.

... it is a frontier of new lands, located only a few days travel time away from the Earth, and built from materials and energy available in space.

... Under the space-colony conditions of virtually unlimited energy and materials resources, a continually rising real income for all colonists is possible— a continuation rather than the arrest of the industrial revolution.

... knowing that the resources of space are so great, we who may be among those first to exploit them can well afford to provide for our less advantaged fellow humans the initial boost that will permit their exploiting these new resources for themselves. Suddenly given a new world market of several hundred billion dollars per year, the first group of nations to build space manufacturing facilities could well afford to divert some fraction of the new profits to providing low-cost energy to nations poor in mineral resources, and to assisting underdeveloped nations by providing them with initial space colonies of their own.

If we use our intelligence and our concern for our fellow human beings in this way, we can, without any sacrifice on our own part, make the next decades a time not of despair but of fulfilled hope, of excitement, and of new opportunity.　　　Gerard O'Neill:
CoEvolution Quarterly
Fall 1975

Dr. Gerard O'Neill, high-energy physicist at Princeton University is the foremost designer and promoter of space colonies in America today. Stewart Brand, founder of the Whole Earth Catalog, *is editor of the* CoEvolution Quarterly *magazine and the book,* Space Colonies.

1975 *Cutaway view of O'Neill's design for space colony of 10,000 inhabitants. Model III would be 1.24 miles diameter, 6.2 miles long.*

... Give your imagination a Space Colony of 1,000,000 inhabitants, each of whom has five acres of land. Know that it's readily possible — maybe inevitable — by 2000 AD ....

O'Neill notes that the ends of the enormous rotating cylinders could be mountain ranges, with the interesting property that as you climb higher your weight decreases. Near the top, at .1 g (1/10 of Earth gravity) you can don wings and take flight. Or you may want to take a long slow plunge into a swimming pool. Or watch someone else's slow-motion splash. At the foot of the mountains you might have a round river, allowing you to canoe downstream several miles past the other two "valleys" and back to your home ....

O'Neill expects that the colonies, once they begin to proliferate, would make themselves as appealing as possible to attract immigrants from Earth. Since the cylinders are big enough to have blue skies and weather, you might design a cylinder pair to have a Hawaiian climate in one and New England in the other, with the usual traffic of surf boards and skis between them (travel in Space is CHEAP — no gravity, no friction)....

I think the voters will be interested enough to approve the requisite $100 billion (one-tenth the cost of Project Independence; 10 times the return in energy alone.) Space Colonies show promise of being able to solve, in order, the Energy Crisis, the Food Crisis, the Arms Race, and the Population Problem ....

Space is part of the wildness in which lies "the preservation of the world"....
Stewart Brand:
CoEvolution Quarterly
Fall 1975
continued

Space Colonies *continued*

We must become citizens of the Universe. The Universe says No to us. We in answer fire a broadside of flesh at it and cry Yes! Other worlds do not live. We will stir them with our limbs. Other parts of the Universe cannot see. We will bring a gift of eyes. Where all is silence, this thing that we call Human will speak . . . .

from Ray Bradbury's Introduction: *Colonies in Space,* T. A. Heppenheimer

The first space community will be economically productive only if talented, hard-working people choose to live in it, either permanently or for periods of several years. It must therefore be much more than a space-station; it must be as earth-like as possible, rich in green growing plants, animals, birds, and the other desirable features of attractive regions on earth.

. . . natural sunshine, a hillside terraced environment, considerable bodies of water for swimming and boating, and an overall population density characteristic of some quite attractive modern communities in the U.S. and in southern France.

From the valley area . . . streams could flow, a ten-minute walk could bring a resident up the hill to a region of much-reduced gravity, where human-powered flight would be easy, sports and ballet could take on a new dimension, and weight would almost disappear. It seems almost a certainty that at such a level a person with a serious heart condition could live far longer than on earth, and that low gravity could greatly ease many of the health problems of advancing age . . . .

Gerard O'Neill

The flying carpet of the Arabian Nights, the seven-leagued boots, the wishing ring, were all evidences of the desire to fly, to travel fast, to diminish space, to remove the obstacle of distance. Along with this went a fairly constant desire to deliver the body from its infirmities, from its early aging, which dries up its powers, and from the diseases that threaten life even in the midst of vigor and youth. The gods may be defined as beings of somewhat more than human stature that have these powers of defying space and time and the cycle of growth and decay . . .

Lewis Mumford: Technics and Civilization

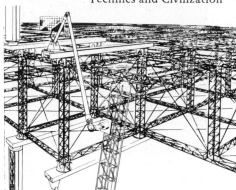

Automated assembler in large-scale space construction.

Colonies Ad Infinitum
The colonies would multiply like cells, giving birth to more colonies ad infinitum, capable, within 60 years, of absorbing the earth's annual population increase. And eventually more, leaving earth with far fewer people than it has now, perhaps just a small resident caretaker force . . . .

San Francisco Chronicle Oct. 5, 1977

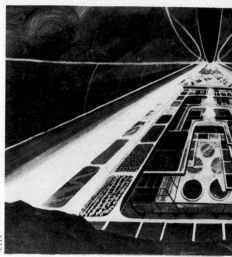

Agriculture for a space community would be carried out in external cylinders or rings, with atmospheres, temperatures, humidity and day-length chosen to match exactly the needs of each type of crop being grown. Because sunshine in free space is available 24 hours per day for 12 months of the year, and because care would be taken not to introduce into the agricultural cylinders the insect pests which have evolved over millennia to attack our crops, agriculture in space could be efficient and predictable, free of the extremes of crop-failure and glut which the terrestrial environment forces on our farmers . . . .

Gerard O'Neill

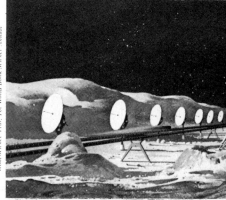

On the moon we could assemble a launching device for the acceleration to escape velocity of lunar surface raw materials . . . .

A collector at escape distance from the moon would accumulate materials, and there, with the full solar energy of free space, they would be processed to form the metals, glass and soil of the first space community. . . .

The machine to transport the lunar material is called a mass driver; it exists only on paper, but it can be designed and built with complete assurance of success because it requires no high-strength materials, no high accelerations or temperatures, and its principles are fully understood.

Gerard O'Neill

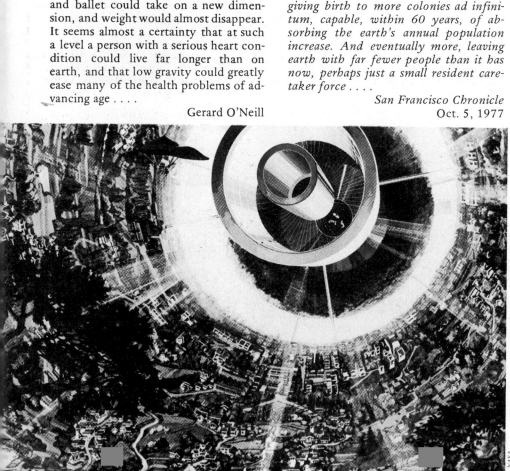

It is entirely reasonable to plan to grow grain in the space farm at a rate of 850 pounds per acre per day. Plants will grow in sand, vermiculite, styrofoam, or nothing at all provided they are supported and receive nutrients and water. Carolyn and Keith Henson, Tucson agriculturalists, propose to support the plants by means of styrofoam boards. The roots would hang below the boards and it would be possible to spray a nutrient solution onto these roots directly . . . .

T. A. Heppenheimer:
Colonies in Space

The largest colonies now foreseeable would probably be formed as cylinders, alternating areas of glass and interior land areas. From those land areas a resident would see a reflected image of the ordinary disc of the sun in the sky, and the sun's image would move across the sky from dawn to dusk as it does on Earth. Within civil engineering limits no greater than those under which our terrestrial bridges and buildings are built, the land area of one cylinder could be as large as 100 square miles. Even a colony of smaller dimensions could be quite attractive . . . .

Gerard O'Neill

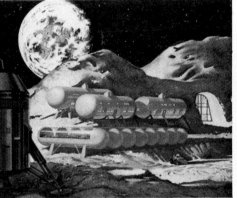

One of the payload canisters will contain a most important item, to be handled with great care: the interim nuclear power plant. To bring a nuclear plant to the sunny moon may be seen by some as akin to bringing oil to the Arabs. There will be much interest in solar power for the lunar base but on the moon, the sun shines only two weeks of the month and nuclear power is available round the calendar . . . .

T. A. Heppenheimer:
Colonies in Space

The quotes on these pages are from Gerard O'Neill are from a lecture given in Spring 1975 at a meeting of the World Future Society in Washington D.C., and from O'Neill's testimony before the Sub-committee on Space Science and Technology, U.S. House of Representatives, July 23, 1975. Both lecture and testimony are reprinted in full in Space Colonies, *ed. Stewart Brand, 1977, Penguin Books.*

Space Clippings

Normal Evolution

"(Humans colonizing space) is just as normal as a child coming out of its mother's womb, gradually learning to stand, then running around on its own legs."

Buckminster Fuller:
Harrowsmith
Sept. 1977

Top NASA scientists are convinced that intelligent life exists in outer space

National Enquirer
Sept. 6, 1977

First space shuttle orbiter Enterprise.

Edwards AFB

Enterprise's two astronauts, describing the third flight of the space shuttle as busy but enjoyable, successfully guided the orbiter to a landing on a dry lake bed yesterday, testing for the first time a sophisticated landing approach system.

San Francisco Chronicle
Sept. 24, 1977

Our New Science

Science is constructing models of reality that are of great beauty as well as amazement, and of great skittishness as well as power, so that our universe now seems arrayed in a new and brilliant dress.

Science has made it a thing of irridescence, a joy to contemplate. Creation is deprived of none of its wonder by the understanding of our new science; with each model of it that we are given, it appears to be only more wonderful . . . .

San Francisco Chronicle
Oct. 16, 1977

Prince Charles to View Final Test Flight of Space Shuttle

Los Angeles Times
Oct. 22, 1977

Military Using Space Shuttle

The Pentagon has quietly begun using NASA's Space Shuttle program as a stepping stone toward building a capability to fight in space, making space warfare a practical possibility for the mid-1980's.

Military space projects, in fact, now make up a significant portion of all NASA's planned shuttle missions. More than 100 of the first 560 shuttle flights will carry US military satellites into orbit.

Milwaukee Journal
Milwaukee, Wisconsin
Oct. 23, 1977
continued

Spherical spaceliner bringing new immigrants to large type colonies as proposed by O'Neill. At right, idyllic country scene inside a large colony as depicted by NASA illustrator Don Davis.

from Science Year, The World Book Science Annual

Clippings *continued*

Space 'Hospital,' Solar Power Plant Envisioned

The nation's space shuttle, now being tested, could someday be used in the development of a "space hospital" or a solar electricity generator, former astronaut Donald (Deke) Slayton said . . . .

Los Angeles Times
Oct. 23, 1977

The Air Force Eyes a Star War

Squat and fat on a Southern California airstrip, the Space Shuttle Orbiter may soon become the prototype of the first American warship to sail in space. The Pentagon has quietly begun using the National Aeronautics and Space Administration's (NASA) new space shuttle program as a steppingstone toward such a vessel, and the growing military interest in space as a potential battlefield presents the nation's civilian space agency with a gloomy and potentially fatal array of problems on the eve of its twentieth birthday . . . .

NASA, now beyond the fast and loose days of the moon race, has been casting around for several years for a mission to pursue during the last twenty years of this century . . . .

One Air Force general summarized the military's view of the situation: "There has never been a transportation medium in the history of man that has not been exploited for economic and military advantage. Space is not going to be an exception."

The Nation
Jan. 7, 1978

War in Space

. . . Military space projects already take up a significant portion of NASA's planned space shuttle missions. More than 100 of the first 560 shuttle flights will carry U.S. military astronauts with satellites and weapons experiments into orbit . . . .

Air Force scientists are already scheduling the first ten military shuttle missions. They will include defense and fleet satellite communication systems; new sophisticated early warning systems against Soviet missiles; laser weapons . . .

Whatever fears others may have that the arms race is being expanded in a new direction, military planners are excited about possible star wars. "I think the space shuttle has been letting a lot of demons out of the cave," one Pentagon scientist told *Aeronautics and Astronautics* recently. "Space is a dandy arena, actually, but you've got to attract strategic war off the planet. The notion of abhorring war in space is just plain wrong."

The Nation
Jan. 14, 1978

Science Fact, Not Science Fiction

. . . By the year 2000 we will be heating our homes with energy beamed directly from the sun, everyone will wear his own personal telephone on his wrist, and television sets will receive 512 channels.

And that, said Jon Michael Smith, director of marketing for the space shuttle at the National Aeronautics and Space Administration, is "science fact," not "science fiction."

"NASA is working today to develop a commercial use for outer space. And we know there is no limit to the things we can do in orbit," he said.

"It's about time we (NASA) opened space up for the American buinessman," Smith said. "We have a very small budget and we just can't take on a project this size alone."

The Evening Bulletin
Philadelphia, Pa.
Jan. 16, 1978

Soviet Nuclear Spy Satellite Disintegrates over Canada

San Francisco Chronicle
Jan. 25, 1978

Death Star, Star Wars

Editorial

When Gulliver visits the flying island of Laputa and its kingdom Balnibarbi, he observes that the inhabitants are abstract thinkers who study only mathematics and music, and fear their world will be imminently destroyed. Scientists and professors seek breakthroughs with disastrous results, yet continue to be subsidized by the rulers of the kingdom. Fields are farmed by mechanical contrivances and nothing grows, the houses are ill-built due to lack of respect for traditional building skills, and professors debate about " . . . the most commodious and effectual ways and means of raising money without grieving the subject." (*Gulliver's Travels,* Jonathan Swift, 1726.)

America's program to colonize space is neither 18th century satire, nor 20th century science fiction. It is real, the early stages of the program have begun, and a smooth, well-orchestrated public relations program is now underway to persuade American taxpayers to pay some enormous bills for the "colonization" of space.

Space colonies, we are being told, will solve the world's food *and* population problems, create new international markets of several hundred billion dollars per year, eliminate the need for mideast oil and domestic nuclear power stations, stabilize world tensions, promote international cooperation, and reduce the prospects of war.

Some features of the program:
— Construction of immense (1¼ miles in diameter, six miles long) metal and glass cylinders in the moon's orbit for 10,000 inhabitants. Advocates envision running streams, grassy hills, and pest-free agriculture.

— Materials for this to be provided by mining the moon. Metals, minerals processed (smelting, etc.) in space with *nuclear power.*
— Solar energy collected on power platforms the size of Manhattan Island, beamed to earth via microwave for conversion into electricity.
— Cost: $100 billion, give or take $50 billion, over a ten year period — Princeton physicist Gerard O'Neill's 1975 estimate.

But let's look a bit closer and ask a few questions:
— Do we ask the rest of the world if it's all right to mine the moon? Does it belong to the United States?
— Topsoil, grassy hills, abundant water? Accomplish in a few decades what took billions of years to evolve naturally on the earth?
— How can we know this program will not have, as with many other brilliant visions when brought from abstract into physical reality, unforseen and disastrous side effects? Soldiers exposed to atomic radiation in the '40's developing cancer 30 years later; the soil fumigant DBCP causing sterility and tumors; polyurethane foam burning like napalm, etc.
— Peaceful use of man-made islands orbiting the earth? How long before the military takes over completely?
— How much is $100 billion? - $500 from every man, woman and child in the U.S. = 2 million new $50,000 homes = 10 million new $10,000/yr. jobs for unemployed people on earth to do useful things. And what cost estimate these days does not have at least 2-300% over runs?

We should question the scientists, technicians and promoters who tell us development of space technology will solve our present problems. Space colonies will never be the peaceful, idyllic settlements now being portrayed, but rather extensions of centralized control, privileges for the few, and instruments of war with immense destructive potential.

We might have learned a thing or two when we got to the moon in 1968. A barren, desolate landscape, devoid of life. Yet on the horizon, a vision of beauty: a cloud-wreathed planet with oceans and mountains, valleys and streams, topsoil and life . . .

It could have been a signal to us to work with what is right here at hand; to find a balance between our needs and the earth's resources; to repair, restore, conserve and rebuild.

It's time to stop wasting precious planetary resources cluttering space and defiling the moon, to stop making 200,000 mile journeys in search of new frontiers to exploit when we could be making better use of what we have right here. It's time for us to get back down to earth. □

Letters

Mud Skyscrapers

Jerry Erbach worked in southern Arabia for two and one-half years as an architect. He writes " . . . The Yemenis still build in the traditional manner, but things are beginning to change and modernization is beginning to take place with disastrous consequences for the architecture . . . ." He sent along this photo of one of the skyscraper castles of the area, " . . . a typical Sana's house, although the style of architecture varies greatly throughout the country . . . ."

To Shelter Publications

. . . an experimental archaeology project that I worked on this past summer. Ten people constructed six structures using materials gathered in the local area. We lived in all these structures for the duration of the project (12 weeks). Our object was to experiment with building houses using post hole patterns found during excavations and using only those materials that the native Indians would have had available to them. We also kept a garden of corn, beans, squash and native tobacco. The houses and garden were built and maintained using stone tools (meaning high energy expenditure on our part but little or no energy expended on Earth's part). The time period we were attempting to replicate was 900-1200 A.D. The site was at Cahokia Mounds in Illinois which boasts the largest Indian earthwork known in North America . . . .

Sincerely,
Steve Koeln
St. Louis, Missouri

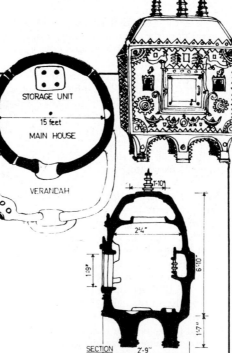

Storage Unit

In many tiny Indian communities. . . these leggy clay storehouses are used for food preparation and leftover dairy products. Since food is looked upon as a divine gift, here in every household this sun dried object gets a privileged position. It invariably dominates the central axis of architectural composition. Its conspicuous exhibition through the street entry to the house charms all visitors by the reflection of light and shade in the embedded mirror bits.

What is striking about them in the present-day-utility is their preservation quality even during inclement weathers of the arid region. The insulation is near perfect. Dairy products like milk, butter, curd and other perishables remain fresh for many days. The interior remains germ free because of the germicidal properties of the basic ingredients.

The raw materials are indigenous clay, dry husk and horse or cowdung; the last two mixed for giving stability and strength to the potbelly shape when it stands on its four legs. The three materials are compounded in secretive proportion (the exact mix ratio is by intuition only) with water and kneaded into stiff dough like well blended paste. The whole mass is then kept moist over three weeks, by which time a craftswoman will produce one perfect container for ready use.

The standard unit is custom made to suit the housewife's hands who is going to own it and use it. . . .

Uttam C. Jain
Bombay, India

Dear friends:

. . . some of the Maori type of dwellings which have been reconstructed here in New Zealand . . . . One family we know well has lived in such a *whare* for two years with no hardship — true, it was in the north of the North Island where the weather is less strenuous. The Maori houses included such features as insulation, double doors to keep out draughts — but their biggest failing was in means of letting out the smoke. Much of their life and their cooking was out of doors — or in cooking shelters. They also made use of excavating down about 3 feet for their houses, giving the warmth of the earth . . . .

I think there is something to be learned — as you do — from native building methods. Corrugated iron roofs on so many of the new houses in the subtropics are an abomination — hellholes in the hot sun; the Maori palm thatch or raupo matting and thatch work well; worst fault is their relatively short life. But if they cost nothing and replacement does not hurt the environment, time is the only thing involved . . . .

Bob Stowell
Diamond Harbour,
New Zealand
October 3, 1974

214

"Bay of Steam"

. . . The use of thermal heating in Iceland was first recorded by Snorri the Saga writer in the 13th century; who had a thermal pool abutting his sod and lava rock home. Ninety-nine percent of the homes in Reykjavik are thermally heated. To tap this resource, 9 bore holes heat the homes of 90,000 residents. Reykjavik means "Bay of Steam". This thermal heat saves thousands of tons of fossil fuel annually. Recently, The Municipal Heating Council and Icelands Department of National Energy ran conduit through-out the downtown mall to keep the Capital's streets and sidewalks warm and snow-free in the colder season. Thermal is used in the Greenhouse industry where tropical fruits are grown 100 miles from the Arctic Circle. The old Viking homes are somewhat reminiscent of Mandan Indian lodges I've visited in North Dakota . . . .

> James P. Miner
> Santa Monica, California
> August 23, 1974

Dear Shelter,

. . . I enclose a few photos of *pallozas* which are ancient circular houses found in the mountains of northern Spain . . . .

> Yours,
> Mark Gimson
> London NW1

Two communal buildings. Pasture used to be orchard. Each village had 100 acres of land and could support 75-100 people. Vegetables, nuts, seeds, fruit and grain grown and eaten.

Spirit Wrestlers

Photos from Robert Wishlaw and J. Bellows of a Spirit Wrestler village, British Columbia, Canada. Occupied 1912-1972

Spirit Wrestlers — pioneers on Canadian Prairies — starting from scratch.

Letters *continued*

Hi,

I've had your book, *Shelter*, for some time now and each time I pick it up to read I find something interesting. It has been a good investment and I congratulate you on doing something which is helpful and enjoyable.

The last time I escaped into *Shelter*, I found your request for material for a *Shelter II* — a great idea and one which I would like to be a part of if what I have to offer is of interest to you.

For the past fifteen years or so I've been either building or remodeling something or other. It is a very important activity in my life. Teaching school is the way I earn a living and it is enjoyable and sometimes even rewarding. But I find that I need something more tangible to make me feel that my stay on Earth is worthwhile. It all started when I wanted a house — not an ordinary house — but one with some character and warmth. I designed a house and then set about making it a reality. Not having any money made that just a little difficult. My efforts to save for a down payment were always trailing behind the rising costs of construction. Without going into the details, I gave up on having the place built and set about doing it myself with various kinds of help. After having talked to builders I felt that if they could do it then I could. The project was a success and I became very courageous.

Following the house there was a tool house which now serves as my pottery. This was followed by the construction of a 2½ car, two story garage with the upstairs finished. This now serves as a guest house and studio. Following this I renovated a barn — a wooded barn — which should have been torn down but I just couldn't face that. It was part of a farm which I bought. The barn now serves as an art gallery in the summer and a great place

to have large parties at other times. The next project was designing and building a 44 cu. ft. gas fired kiln and building to house it. Much to my surprise it worked the first time.

But, last Summer I did one of the most enjoyable building projects of my "career". I built a small, two story cabin (12x24) up in the mountains on some land I had purchased the year before. There is no electricity up there and I don't want any either. So, the work had to be done without the aid of power tools. Some of the boards were cut at home and hauled up there but a good deal of the cutting had to be done by hand. The building sets on ten pilings, 24" deep and partly filled with gravel. All this had to be done by hand, the gravel wheeled up the hill because the old '61 couldn't make it. That was the hardest part of the project. The outside is all done (I worked at it for about 2 months) and am anxiously awaiting next Spring so that I can finish the interior. I've done most of the work alone asking for help only when I realized the risk of doing a couple of things alone. It has been a great experience for me up in the mountains alone finding my limits; the quiet except for two streams that create a stereo sound and the three ravens that kept watch over my activities. The days were long and exhausting but the rewards were great.

On one of the days I took a nice bottle of wine with me and at noon I sat beside a stream which is part of a lovely natural Zen garden and I drank a toast to the trees and the streams and the ravens and me on my 53rd birthday and to the fact that I was a woman doing it MY way.

Again, thank you for your book; it's nice to know that there are others out there stubborn enough to do things their way.

Mary M. Carrabba
Spokane, Washington

Carrabba house

Dear Shelter,

This *Shelter* book we have has been inspiration for us. One fantastic 16 sided, almost round bush timber home styled after the Mandan lodge hit the dirt in a big rain and wind before we finished it. Now we live in a bush house . . . with hessian burlap and lime walls. Very cheap and so far so good. We painted it on inside and out and the hessian is nice and tight and white after drying, although it looks grey when wet at first.

A surprise bonus: unpainted sections give screened windows.

Jim Savage & Lynn Halpern
Jerramungup, Western Australia
P.S. Since we built it, older visitors tell us most people in the area started out with a place like ours; bush timber and hessian and lime.

Shelter People:

Have been reading the *Shelter* book and thought I'd tell you about our house. Basically, it is a log hexagon, walls 12' on the inside. Hex, partly because we like the shape and also so we could use shorter logs and get more room with more than 4 sides. All the logs were either dead standing or leftovers from logging operations, cut *no* living trees. We put in the window and door frames as we went and were able to use peeled poles given away by a pole company for short logs (such as between a window and a corner.) We tore down an old house and got enough used lumber (beautifully seasoned) for all of the frames, the sleeping loft floor, and other necessities. By exchanging labor with our neighbor who has a sawmill, we received 18 2" x 6"s for floor joists. Foundation is old concrete 12" pipes, filled with cement. The subfloor is mill ends, free from a local mill. The windows we found in demolished buildings or were given to us. The roof had us stumped, 'til out of the blue, *Shelter* was sent to us. So our idea came from there, we used 8" logs, 16' long butted and spiked to a kingpost

We've been told that the weight of the roof will force the walls out and the house will collapse. We know nothing about stresses, etc., but it seems pretty strong, has a fairly steep pitch (8' rise over 12' run). We built the roof so you can see the poles inside, found some beautiful shiplap in an old granary for paneling the roof. Also put 3½" fiberglass insulation on roof. Only expenses were: insulation, tarpaper, nails and sundry. Old nails are far superior to new ones. Also we used handtools, no electricity and no chainsaw.

Peace
Vicki & Allen Armstrong
Colville, Washington

Hello,

. . . I spent 9 months living and working in Holland, mostly Amsterdam in the last year. Over there, its called black work when you don't go tell the gov't. about it. Sure was a pleasure, most of the builidngs I've been working on are 2-300 years old and still going strong. Brick on the outside, wood floors. I was living in one of the more together cracked houses, cracking -- squatting. Its a large warehouse, in the center of Amsterdam, built in the 1600s, preserved by the gov't. as a monument house, owned by I.T.&T., but owing to the neighbors, they couldn't do much with it, and so several years back, it got occupied, and passed to us last October. We've been cherry-ing it out, restored the front of it, got a garden out back, sound studio, garage, workshops, lights/heat/hot water, and the silent approval of the neighbors earned thru cleanliness, friendliness, and quietness after 10. Neighborhood political power is still alive in Holland some, altho a monstrosity known as the Metro, is inching its way thru town -- it being a subway, built above ground and then sunk. If you've not been there, the city is built on sand & water, and I never met a Dutchman who thought the Metro would work . . . .

David Handel
West Winfield, New York

Greetings:

My name is Bob and I will be building my home in the spring of '76 . . . .

I've gathered information from "The Whole Earth Cat.", "Shelter", "Illustrated House Building", "The Owner-Built Home", etc., etc. . . . and have been "grubin'" materials like windows, doors, tools and wood from people's garbage. I live presently in the suburbs of Cleveland, ugh! but will be moving to my land. I bought 2 acres last year with money I made from teaching guitar. I'm 19 years old . . . .

Thanks,
Bob Belmont
Arkport, N. Y.

Football house in California — from Jersey Devil, Blawenburg, N.J.

Dear friends,

. . . we have a cabin on 20 acres in the Bay of Islands with no electricity. The Aladdin lamp can be used to cook on if you are very careful — suspend the pot on a frame or hang from wire from ceiling beam over the chimney

Dear Shelter,

. . . The house in the enclosed picture started out in 1970 as a 13 x 24 one room cabin which was to have been strictly temporary while we built our permanent home a few hundred yards away. However we have slowly chainsawed openings into three of the four walls to add an upstairs, downstairs, bathroom, bedrooms and meditation room. This summer we'll be dismantling a 60 x 40 square hewn cedar barn and moving the logs near our home. The summer of '76 will see my trusty chainsaw carve out an opening in

the only wall without an addition. Here we'll reassemble the cedar logs for a living/craft room, veranda and add our nine sided silo to one corner. We're still not sure what the silo will be, but that's no problem since we've rarely known what we've been doing until suddenly we flashed on what we had done.

We've spent less than $5000 including tools. It's all log, cedar, plaster, rock and cement . . . .

Valajan Heckrodt
Enderby, B. C.

being careful not to get it too close. The Aladdin gives tremendous amount of concentrated heat. We used one for five years in our Vermont homestead and often kept one mantle for half a year. We also had a 6 volt windcharger for a 15 watt bulb in the cow stable because we did not like kerosene lanterns in the barn with all the hay. We just used it to charge car batteries as needed. But by far the best both for individual and the planet is to use the natural daylight; arrange to do your talking, music, in the hours after sunset when no light or only the most frugal is needed . . . .

. . . I am collecting stories on the cruising experiences of people in open sailing dinghies — under 18 foot . . . to show that expensive gear and boats are not needed for extended coastal voyages. (We have an International 14, not ideal in some ways, but, sailed conservatively, it will do the job and has the advantage of extra speed to get you out of trouble when needed.) Uffa Fox crossed English Channel in one — Southampton to Le Havre; other Englishmen have made notable small boat voyages to Norway, Iceland — what about Americans? This is one of the last places for direct contact with relatively uncontaminated nature and it need not cost more than the price of a second-hand 14 to 18 foot sailboat. Safety is essential but with a light boat you can manage to pull it up on most beaches — even rocks can be handled if you carry the right gear — then you wait out the bad weather . . . .

Bob Stowell
Diamond Harbour, N.Z.

Dear Brother or Sister: we enjoyed *Shelter* very much, anyone I know who is into carpentry or simple home building all raved about it. We were glad you mentioned Te Rangi Hiroa's book *Arts & Crafts of Hawaii.* This is one book you can hardly get at the Maui Library even tho they have ten copies. It's now printed up in chapters and all are a storehouse of information, especially *Houses II* and *Food I.* Living on the Islands is definitely different than Santa Cruz Mountains where my wife and I lived 3 years ago. We have just finished our floor, roof, and we have our tent set under it. The foundation is about 4 feet off the ground. We get lots of rain at times, but when the sun comes out it's unbelievable. We live in a huge valley with waterfalls and streams all around. Building codes aren't heavy. Everyone we know has a privy. No elec or plumbing; we use propane and rain water collected by our homes. Most people build shed type homes altho I know of one dome home with tin can shingles *that doesn't leak.* Rain barrels are up higher than sink for gravity feed. It's warm year around so not too many people have heat except up country, and gardening is year around. We're growing bananas and papayas besides our huge garden, seems like there is always something to do. Some of the older buildings are outstanding and I don't know exactly what they are called, maybe Victorian Hawaiian . . . .

Michael Graham
Makawao, Hawaiian Islands

Viking ship — from Carla Bombere, Hannover, Germany

Dear People:

...I do not feel that you have produced a book which will have very much influence. You are too far out for the person who can only think of a trailer as an answer to low cost housing needs. (in 1972 80% of all new housing for native Vermonters was trailers). You made no mention of how to deal with building codes, how to deal with inspectors. From the point of view of provision of services — electricity, transportation, schools, it is more reasonable for people to cluster in villages than live in rural isolation....

Minor criticisms

1. Too many pictures — some of them not at all clear
2. Repetition: and inclusion of irrelevance If I buy a book on shelter and find Indian stories I wonder how much of the rest is really relevant — like reading a newspaper story you know to be untrue makes you doubt the whole paper.

Best thing

The change from domes to natural local materials. It is refreshing to see someone ready to admit in print that they were going off in the wrong direction....

...I guess my impatience comes from seeing so much rural poverty in housing — and knowing that people do not have to ape those with much more money in order to house themselves better. Post war England had housing co-operatives — a self help method of dealing with a housing shortage... anyway there's lots more to put in your next book.

Sheila J. Dusenbury
Dorset, Vermont

Dear Shelter:

Ten years ago reading Fuller, I thought it was housing for after the next war. Ten years later — we have land — not wanting an old time heavy house, not having that kind of loot anyway, having seen no more than 2 really 20th cent. houses in all this time, not having the loot for that either —

or the conviction that that was ours to do — meeting a dome man with an easy solution via foam and a handsome expanded relaxed dome design...so we looked forward to continuing the skyscape we'd loved in an old old NYC apart. and accepting that it was already after the next war — inherent in foam-land. A downward acceptance, finally refusal. No, it *is* dwelling space for the moon, not *yet* practical and proper for Earth, such simplistic centering not *yet* sufficient for earth's native, a relief to still have time to live, not just be preserved. Step by step... For the moment we are hoping on acquiring old freight cars! How comfortable that feels. One thing will lead to another. Glad to be hearing of your journey.

May Asher
New York, N.Y.

Dear Shelter People:

... I would like to make a correction about Eucalyptus. What scant info you included on page 62 of *Shelter* only left out well over 400 of the Eucalyptus -- there are some 450 different varieties with at least one available for every conceivable purpose....

Yours,
G. Woodside
Christchurch, N.Z.

P.S. Checks in a whole pole do not reduce strength.

dear shelter people/ grabbed a copy of your book and it sparked off a stream of ideas and revelations etc so having just built my house i thought i'd write you.

what i'd like to do is contribute a section on *practical building* and building in the antipodes . . yr sea climate and availability of materials make shelter not particularly practical to antipodean builders. things like shingles are unavailable in our part of the world and building restrictions are pretty tight... so anywhat here is what we did...we found an old house in the city which was due for demolition. we offered the owner to demol-

ish it for free (which saves him money) and then with the demolition material we built a 15 x 19 cabin in the back yard of the house we were demolishing; we lived in the part demolished house during building.

having resited our cabin we then moved the remainder of the demolition material to our land and started building the second section (which was the same size as the first cabin) at a right angle to the original cabin. the total floor area was 570 square feet with a loft covering one third of the second cabin for sleeping and storage. the house is timber framed with fibre-light cladding and floored and lined in kauri boxing. we used a septic tank for sewerage and our hot water is supplied by solar heating and water by a windpowered pump. the total cost was $350.00 incidently the house is really just a development of the cabin you showed on page 41 of your book....

love and good happenings, arohanui
jon adams
roskill, south auckland, N. Z.

Shelter:

This is the most beautiful barn in the world. Great Smoky Mountains National Park, reconstruction of an original pioneer structure. Impeccably detailed. Mortise & tenon construction....

Jim Toohey
Knoxville, Tennessee

Shelter editor,

My wife and I are living in a place we've built that might be of interest to Shelter II readers who live in Northern climates. It combines aspects of both the eskimo sod igloo and the esthetics of a log cabin. Essentially a log cabin with a double door antiway ("cold door") built in, it is dug into the forest floor about 2' with sod block walls built up to window level on the outside of the logs on three sides. One side has sod outside the logs to the eaves (this is on the stove side and acts as a heat resevoir — radiating heat back into the cabin throughout a fireless night.) The roof is of closely laid poles, visqueen, and a heavy covering of moss and sod. The house, except plastic and nails, is of native materials (primarily spruce) and is heated on an armload of wood for 24 hrs. of deep subzero cold. We've also got a good modification of a 30 gal. drum stove with a 10 gal. "stack robber" oven. Have made all furniture, wood hinges etc.

Seems like a lot of ego tripping but this has turned out to be a happy home and a sturdy one. We're 45 mi. from others (Eskimo and some whites) so a reliable shelter is a must....

Jack Hebert
Ambler, Alaska

Dixon house

Dear Shelter:

. . . In the past summer I lived in the coastal range of Humbolt County just north of Eureka. I was able to make great use of the natural resources there and was able to construct a shelter almost entirely from the natural surroundings. There, as you well know the refuse of the logging operations abounds in catastrophic quantities. I constructed a 16 foot Icosahedron cantilevered on top of a 6-8 foot diameter redwood stump. I used 3-4 inch diameter douglas fir poles for the skeleton and sheathed the structure with shakes hand split from "left behind" shake bolts and did the interior with 1 x 6 redwood salvaged from an old mill. Glass came from the same mill and was cut and mounted in frames made to fit the triangular sides of the structure. The only expense was the 30 lb. felt paper and 4 mil. poly. used for underlayment for the exterior shakes and for the central skylight . . . .

John Dixon
Pescadero, California

On a trip through northern Ankara, Turkey, we saw this house with its balustrade of recycled truck tyres. It was on the edge of a dense forest where timber is abundant. The village had a Roman bath with natural springs still in use. Four men from Seyhamami looked at us with pride, assuming we were taking their picture. When we indicated we were interested in the balustrade, the one who built it said: "It is trivial, but it does its job," then added, "but it's beautiful, ain't it?"

Suba Ozkan, Ankara, Turkey
Photo: Y. Oymak

Bibliography

Adobe – Build It Yourself. Paul Graham McHenry, Jr. The University of Arizona Press, Tucson, Arizona. 1973.

Step-by-step instructions on building an adobe home.

America's Forgotten Architecture. Tony P. Wrenn and Elizabeth D. Mulloy. National Trust For Historic Preservation, Pantheon Books, New York, N.Y. 1976.

Over 400 photographs of unique and mostly unknown American architecture.

Architecture for the Poor. Hassan Fathy. The University of Chicago Press, Chicago, Illinois. 1973.

Moslem village built by craftsmen trained on site in 1940's by architect Fathy.

L'Art de Restaurer une Maison Paysanne. Roger Fischer. Librairie Hachette, La Cité des Livres de Los Angeles, 2306 Westwood Blvd., Los Angeles, Ca. 90064. 1966.

Unique book of rural French houses; hundreds of photos; text in French.

Barns, Sheds and Outbuildings. Ed. by Byron D. Halsted. The Stephen Greene Press, Brattleboro, Vermont. 1977.

Reprint of the original published in 1881; straight-forward country construction.

Built to Last: A Handbook on Recycling Old Buildings. Massachusetts Dept. of Community Affairs. The Preservation Press, Washington, D. C. 1977.

A record, with photos, costs and other data of rehabilitating old buildings in Massachusetts.

Cabinetmaking and Millwork. John L. Feirer. Chas. A. Bennett Co., Inc., Peoria, Ill. 1970.

Materials, tools, machines, techniques for cabinet work, furniture, finish carpentry.

The Complete Book of Woodwork. Charles H. Hayward. Drake Publishers Inc., New York, N.Y. 1974.

Excellent general introduction to tools, woodworking and cabinetry.

Don't Go Buy Appearances. George Hoffman. Woodward Books, Corte Madera, Calif. 1978.

House inspection guidelines. Invaluable for prospective home buyers.

Dwelling Construction Under the Uniform Building Code. International Conference of Building Officials, 5360 South Workman Mill Road, Whittier, Calif. 90601. 1976.

Abbreviated, simplified version of Uniform Building Code.

Elements for Self-Knowledge – Towards a True Architecture. Aris Konstantinidis. Books Endohora Co., Ltd., 62 Solonos Str., Athens (135), Greece. 1975.

Stunning account of rural Greek architecture with superb photos (some in color) commentary and drawings by the author.

The English Gypsy Caravan. Denis E. Harvey and C. H. Ward-Jackson. David and Charles, Ltd., South Devon House, Newton Abbot, England. 1972.

History of the English gypsy horse-drawn caravan and its design and decoration.

The English Sunrise. Brian Rice and Tony Evans, Flash Books, Quick Fox, Inc., New York, N.Y. 1973.

A beautiful small book of elegant color photographs.

Fundamentals of Carpentry – Practical Construction. Walter E. Durbahn and Elmer W. Sundberg. American Technical Society, Chicago, Illinois. 1977.

Comprehensive, clear, intelligent text on general carpentry construction; used in most union carpentry apprentice training programs in the U.S.

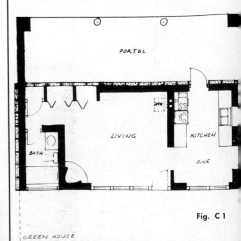

Hand-Hewn: The Art of Building Your Own Cabin. William C. Leitch. Chronicle Books, San Francisco, Calif. 1976.

Practical manual, with good photos, of building log cabins.

The Healthy House. Ken Kern. Owner-Builder Publications, Box 550, Oakhurst, Calif. 93644.

Interesting ideas on radiation, light, sound, heat, air, water flows through a house from the author of *The Owner-Built Home.*

Fig. C 1

Homegrown Sundwellings. Peter Van Dresser. The Lightning Tree, Santa Fe, New Mexico. 1977.

Thoughtful, well-balanced book on low-cost, low-technology owner-built dwellings, and solar heating, mostly in the American southwest.

The Home Owner Handbook of Plumbing and Heating. Richard Day and Henry Clark. Crown Publishers, New York, N.Y. 1974.

Straight-forward, clear plumbing guide.

How To Build A Low-Cost House of Stone.
Lewis and Sharon Watson. Stonehouse Publications, Sweet, Idaho. 1978.
Practical, charming account of building a low-cost stone house.

The Old House Catalogue. Lawrence Grow. Main Street/Universe Books, New York, N.Y. 1977.
2500 products and services for restoring, decorating and furnishing older houses.

The Old-House Journal. Brooklyn, N.Y. Monthly.
Magazine of restoration and maintenance techniques for older houses.

The Old-House Journal Catalog. The Old-House Journal. Yearly.
Buyer's guide for products, materials and services for restoration and decoration of older homes.

Other Homes and Garbage. Jim Leckie, Gil Masters, Harry Whitehouse, Lily Young. Sierra Club Books, San Francisco, California. 1975.
Good source book on alternative energy, insulation, wastes and water supply by four Stanford engineers.

Peace of Mind in Earthquake Country. Peter Yanev. Chronicle Books, San Francisco, California. 1974.
History and causes of earthquakes; what the home buyer should look for; what the homeowner can do.

The Pine Furniture of Early New England.
Russell Hawes Kettell. Dover Publications, Inc., New York, N.Y. 1956.
Over 200 examples of plain, simple early American furniture with some working drawings.

Plants, People, and Environmental Quality.
Gary O. Robinette. Superintendent of Documents, U.S. Government Printing Office. 1972.
Landscaping for privacy, windbreak, cooling, and architectural use. A simple, clear book for the non-professional.

Reader's Digest Complete Do-It-Yourself Manual. Reader's Digest, Pleasantville, N.Y. 1973.
600 pages of clear instructions on building, maintaining, repairing, plumbing, wiring, tools, etc.

The Rock Is My Home. Werner Blaser. Van Nostrand Reinhold Company. 1977.
Dramatic photos of unique stone masonry and houses in Ireland, France and Italy; text in German, French and English.

Shelter. Shelter Publications, Bolinas, Calif. 1973.
Buildings from around the world, construction drawings of sheds and small utility buildings, dome history and review, interviews.

The Timber Framing Book. Stewart Elliot, Eugenie Wallas. Housesmiths Press, York, Maine. 1977.
A good manual on mortise and tenon construction by a group in Maine who build timber frame structures.

Twentieth Century Engineering. The Museum of Modern Art, New York, N.Y. 1964.
The art of engineering: dams, silos, grain elevators, bridges, large buildings.

Victorian Architecture. A. J. Bicknell and W. T. Comstock. American Life Foundation, Watkins Glen, N.Y. 1977.
Fine architectural renderings of Victorian homes in the 1800's.

Villages and Towns: Mediterranean Sea.
Yukio Futugawa. A.D.A. Edita, Tokyo, Japan. 1976.
Beautiful black and white, color photographs of indigenous buildings of the Mediterranean.

Wood-Frame House Construction. L. O. Anderson and O. C. Heyer. U.S. Dept. of Agriculture. 1955.
Very good low cost book on building the stud-frame house.

Woodward's Country Homes. Geo. E. Woodward. The American Life Foundation, Watkins Glen, N.Y. 1977.
Paperback reprint of an 1865 work on rural architecture; fine early American designs and practical information.

Working. Studs Terkel. Avon Books, New York, N.Y. 1975.
Real people talking about real work in America.

Credits

Editor
Lloyd Kahn, Jr.
Art Director
Bob Easton
Contributing Editors (Staff)
Ned Cherry
Bob Easton
Michael Gaspers
Contributing Editors (Field)
Ian Davis
Suha Ozkan
Ole Wik
Burton Wilson
D. Stafford Willard
Consulting Editors
Lesley Creed
Marjorie Jacobs
Rodney Krieger
Herbert Wimmer
Associate Art Director
Jane Krivich
Contributing Artist
Bonnie Russell
Production Staff
Lesley Creed
Richard Fernau
Susan Friedland
Ian Hogan
Marjorie Jacobs
Sara Schrom
Patricia M. Tassa
With Help From
Sally Alexander
Allen's Press Clippings Bureau,
    San Francisco, Calif.
Architectural Association,
    London
Jim Axley
Nina Bellak
Elliot Buchdrucker
Max Croshell
Signe Drury
Jack Fulton
Frank Goad
Winston Grant
Vahé Guzelimian
Bob Hodges
Marie-Christine Kollock
Paula Koolkin
Carolyn Lumsden
Charlotte Leon Mayerson
Phillip Mathews
Rahima Middleton
Richard Niebeck
Louise Pepper
Aubrey and Helena Sheiham
Herman Spaeth
John Stein
Larry D. Thompson
John Welles
Minor Wilson
Barry Wodell

Cover photo
Jack Fulton. Model home in building museum, Dar es Salaam, Tanzania, Africa.
Typesetting
IBM Electronic Selectric Composer
Type face
Journal Roman
Photo printing
General Graphics,
    San Francisco, Calif. (approx. 15% of prints in book).
Color separations
Paddy McLennan, Focus Four,
    Belmont, Calif.
Photostats
Joseph Fay, Marinstat,
    Mill Valley, Calif.
Black and white line-shot reductions
Polaroid MP-3 Land Camera
Printing
Sunset-Recorder Press, San Francisco, Calif.; arranged by Mo Shallat; Director of Operations: Charles Gray; Plant Manager: Jack Parish; Production Co-ordinator: Grat English; Foreman, Prep. Dept.: Stan Hovitz; printing on Harris-Cottrell 700 web offset-M heat set press; hardbound covers by Vince Mullins, Cardoza-James Binding Company, San Francisco, Calif.
Paper
Body of book is 50 lb. Sonoma Gloss

To Order Books By Mail
Shelter, 176 pp, 1973.
    U.S. & Canada, surface mail: $6.50 postpaid. Calif. residents add 6% sales tax. Air mail, U.S. & Canada: $9.00 postpaid.
Shelter II, 224 pp, 1978.
    U.S. & Canada, surface mail: $10.00 postpaid. Calif. residents add 6% sales tax. Air mail, U.S. & Canada: $12.00 postpaid.
    Order from:
    Home Book Service
    P. O. Box 650
    Bolinas, CA 94924
    Other countries:
    Write for prices.
    Hardbound editions:
    Write for prices.
    Bookstores:
    Contact Random House, 457 Hahn Rd., Westminster, Md. 21157

Shelter III

Shelter III will be published when enough material is ready. It will include: indigenous builders; small houses and barns in North America; continuation of small house design and construction; kitchens, bathrooms, work areas, etc.; builders in the cities; rebuilding; housing for the elderly; mobile homes and factory-built housing; maintenance, wrecking and repair; stories and experiences.

If you have something to contribute please contact us now with preliminary information. Contributors are paid on the basis of material used in the completed book.

We would like to hear from:

Builders:
— construction ideas, techniques, tips.
— drawings, photos of houses, buildings you like.
— building experiences and stories.
— comments and criticism on pp. 74-135 of this book.

Designers, architects:
— small house designs, floor plans.
— drawings, photos of houses, buildings you like.
— comments and criticism on pp. 74-104 of this book.

Photographers, travelers:
— small houses and barns of North America.
— indigenous buildings from around the world.
— interiors; kitchens, bedrooms, bathrooms, etc.
— photos similar to those in this book.

Rehabilitators, wreckers:
— accounts of rehab work.
— wrecking, salvage, rebuilding techniques, tips, stories.

Shelter Publications Mailing List
To be on our mailing list for future publications, send a standard postcard with your name and address in the upper left hand corner. We would also appreciate comments and criticisms on *Shelter*.

Shelter Publications
P.O. Box 279
Bolinas, Calif. 94924 USA

Photo Credits

. A. Germen
4. Left top, bot.: James P. Warfield; top right: Jack Fulton; bot. right: Dan Drown.
5. Jack Fulton.
6-9. Paul Marchant.
10-11. Anders Grum.
12-13. Jack Fulton.
14. Left: OXFAM; right: D. Henrioud, United Nations High Commissioner for Relief (UNHCR).
15. Top: D. Henrioud, UNHCR; others: UNHCR.
16. Top and mid.: Ian Davis; bottom: Jean Mohr, UNHCR.
17. Ian Davis.
18-19. Ajit Mangar.
20-21. All James P. Warfield except top rt.: Richard Sexton.
22. Koji Yagi.
23. D. Stafford Woolard.
24-25. D. Stafford Woolard.
26-27. D. Stafford Woolard.
30-31. Suha Ozkan.
32-33. Suha Ozkan.
34. Koji Yagi.
35. Top: S. Sahin; bot. left: Koji Yagi; bot. right: S. Sahin.
36-39. Aris Konstantinidis.
40-41. John Hamilton Doyle.
42-43. Jos Le Doare.
44-45. Roger Fischer.
48. Lloyd Kahn, Jr. (L.K.)
49. Denis Harvey.
50. Smithsonian Institution, National Anthropological Archives.
51. Top: Milwaukee Public Museum; bot: Bob Easton.
52-53. Milwaukee Public Museum.
54-55. Solomon D. Butcher Collection, Nebraska State Historical Society.
56-59. Burton Wilson.
60-61. All L.K. except top right: Jack Fulton.
62-63. L.K.
64-69. L.K.
70-73. from *Bungalows* by Henry H. Saylor.
74. L.K.
75. from *Bungalows* by Henry H. Saylor.
76. Vahé Guzelimian.
77. Samuel Chamberlain.
79. L.K.
. L.K.
. Manya Wik.
. L.K.
. Top: Jack Litrell; bot: L.K.

92. Top, bot. right: Burton Wilson; mid: Minor Wilson; bot. left: L.K.
94. Left: L.K.; others: Burton Wilson.
95. Burton Wilson
96. Bot. left: Samuel Chamberlain, *A Small House in the Sun;* bot. right: Burton Wilson.
97. Burton Wilson.
98. Bot. left: Jack Litrell; others: L.K.
100. L.K.
101. Burton Wilson.
107. L.K.
112, 113. Vahé Guzelimian.
117. Vahé Guzelimian.
120. Minor Wilson.
121. Vahé Guzelimian.
122. Minor Wilson.
126-27. L.K.
128. L.K.
131. Maryann Hcimsath, *Pioneer Texas Buildings.*
136-7. Mark Phelps, Burton Wilson.
138. Top: American Plywood Assn.; bot: A. J. Garr.

139. From Peter Yanev.
141. P. G. McHenry.
142. Top left: *Bambu;* top right: S.F. Chronicle Foreign Service.
144-45. Manya Wik.
146-47. Michael Gesinger.
148. Bill Namaste.
149. Peter & Connie Lundell.
150-51. Lewis & Sharon Watson.
152-55. L.K.
156-57. Ruth & Jan Wheeler.
158-59. Cascade Shelter Co-op.
160-61. Robert Stowell.
162-63. Ian Ingersoll.
164. Top left: L.K.; bot. left: Jack Litrell; top & bot. right: Jack Fulton.
165. All L.K. except bot. right: Jack Fulton.
166-69. Manya Wik.
170-71. George & Nell Abernathy.
172. All Jack Fulton except bot. left: L.K.
173. Bot. left: Jack Fulton; others: L.K.
174. Ron Diamond.

175. Top: Homeowner's Rehab; middle: Ned Cherry; bottom: Mark Haven.
176-81. Homeowner's Rehab.
182-87. Mark Haven, except where indicated.
188-89. Hugo Schuit.
190-91. Andrew Howarth.
194. L.K.
195. Cheryl Morgan.
196-97. Cheryl Morgan.
198-99. George Hoffman.
200. Western Hemisphere, Ltd.
203. Doug Lehman.
204. Top: A. Gunoven; bottom: M. Asatekin; middle: C. Yetken.
205. C Yetken.
206. L.K.
219. Bottom: Y. Oymak.
223-24. L.K.

Drawing Credits
(Those not indicated on pages)
6. Bonnie Russell.
9. Bob Easton (Susan Friedland, inking).
10. Anders Grum.
30-31. Suha Ozkan.
39. Aris Konstantinidis.
40-41. John Hamilton Doyle.
49. Denis E. Harvey.
50-53. Bob Easton.
65. Bonnie Russell.
75-85. Bob Easton.
87. Bonnie Russell.
88. adobe house: *American Building 2: The Environmental Forces That Shape It;* Egyptian rural dwelling: *The Habitation of Man in All Ages.*
88-101. Bob Easton (Susan Friedland, inking).
108-110. From *Village and Farm Cottages,* Library of Victorian Culture.
111. Bonnie Russell.
112-125. Ian Hogan, Bob Easton (Susan Friedland, inking).
126-134. Bob Easton (Susan Friedland, inking).
139. Peter Yanev.
140-141. P. G. McHenry.
142-143. From *Bambu* and *Bamboo as a Building Material*
147. Don Gesinger.
150. Ian Hogan.
191. David Conover.
207-208. J. J. Grandville.

Index

Windmill in France